ERVAN GARRISON

A HISTORY OF ENGINEERING AND TECHNOLOGY

Artful Methods

CRC Press
Boca Raton Ann Arbor Boston London

Library of Congress Cataloging-in-Publication Data

Garrison, Ervan G.
   A history of engineering and technology : artful methods /author,
Ervan Garrison.
      p.   cm.
   Includes bibliographical references and index.
   ISBN 0-8493-8836-8
   1. Engineering--History.   I. Title.
TA15.G37   1991
620′.009--dc20
                                                  91-8782
                                                  CIP

Direct all inquiries to CRC Press, Inc., 2000 Corporate Blvd., N.W., Boca Raton, Florida 33431.

© 1991 by CRC Press, Inc.

International Standard Book Number 0-8493-8836-8

Library of Congress Card Number 91-8782
Printed in the United States of America   3 4 5 6 7 8 9 0

PREFACE

"Firmness, utility, pleasant appearance..." — Vitruvius

An Artful Method was the classical Greek description of technology. The Greek mechanical engineer or mechanaomai, in his craft, was seen as "creating or contriving a deception." The great Hellenistic engineer, Archimedes, created "artful deceptions" such as cleverly contrived gearing to launch large ships. For the title of this book it is appropriate as engineering is both an art and a method.

Engineering is one of the oldest examples of applied art—a unique union of specific need and specific design in a process that yields an engineered work or product. The role of engineering in providing for mankind's material needs is as old as civilization and had its origin in the non-literate ages of man's antiquity. Archaeology constantly provides us with newly discovered examples of man's early ability to provide for his own and society's well-being through the creative act of design.

Man has always been a creator. It is this ability to convert creative thought into substantive works that differentiates him from other organisms. When this faculty developed is known only approximately and is inferred from the fortuitously preserved remains of these first works. The earliest use of fire occurred in middle Pleistocene Africa ($\sim$2 million years ago), along with the intentional manufacture of the first simple tools struck from stone. These simple but powerful events are early demonstrations of the human brain's ability to conceptualize and produce a wholly different form of phenomena from another. It is this perception of causality and its execution into form that is the seminal aspect of design art. Engineering has always shared this fundamental aspect with other creative processes.

The production of a different state of energy, a new and useful form of material are not uniquely engineering acts. They are, however, necessary and sufficient to engineered design wherein energy and tools serve the designer in the creation of some work that in turn serves society's needs. This aspect of engineered design serves both society and the artisan by satisfying the common good and the individual's need for a role. It was only when society itself evolved to require specialists, such as those in agriculture, building, warfare and other cultural roles, did the first engineers take on a distinct and separate status.

The terminology for these early practitioners was clearly defined within the culture in which they existed, as are our names for specialists of any kind. In other words "we are what we do" to paraphrase a recent aphorism. More specifically, a culture or society classifies a member according to his or her role within it. Normally, the term for that role incorporates some descriptive element into the definition of it. Terms for "engineers" or persons who utilize elements associated with the modern perception of engineering have come down to us from past cultures. They typically follow this rough etiological rule of defining the actor by reference to the action carried out, e.g., the Mesopotamian term *batu* simply means "builder".

Today's engineer uses his art as craft and science merged to execute some idea of design. In the description of engineering we hear such terms as "engineering science", "applied science", and "technology" used almost interchangeably. Modern Western nations, as a rule, distinguish scientists from engineers and from architects. Why?

On the distinction between scientist and engineer, Charles H. Holbrow, Professor of Physics, Colgate University offers this: "the scientist is his own client, chooses his own problems, doesn't have to worry about producing a useable end product and generally isn't concerned with long-term efficiency or effectiveness."[1] The distinction between architect

---

[1]   Holbrow, C. H., Russel, A. M., and Sutton, G. F., Eds. *Space Colonization: Technology and the Liberal Arts*. American Institute of Physics (1986).

and engineer is less easily drawn, particularly in the area of civil engineering. Through the Renaissance the engineer was the builder and vice versa, be it Gothic structures of the Middle Ages or aqueducts for the Rome of antiquity. It is only with the modern era that we see a clear distinction in what engineers and specialists from other fields such as architecture and science do. Blurring still occurs today, particularly with structural engineering and architecture.

Robert Maillart's work in concrete bridge design clearly points out this modern convergence of structural ''art'' with architecture. Writers such as Louis Kahn[2] attempt to differentiate the two by saying ''architecture is the thoughtful making of space'' and Elizabeth Mock says: ''Most engineers' bridges cut through space...(the) architect's...give(s) space shape and meaning.''[3]

One can look at Swiss engineer Christian Menn's 1980 Ganter Bridge and easily reject these views. The dichotomy still exists due, in large part, to the Swiss-born French architect Le Corbusier. In *Towards a New Architecture*[4] he proposes this polemical view of engineering and architecture:

> The engineer, inspired by the Law of Economy and governed by mathematical calculation, puts in accord with universal law. He achieves harmony. The Architect, by his arrangement of forms, realizes an order which is a pure creation of his spirit...it is then that we experience the sense of beauty.[4]

According to this distinction engineering achieves harmony, architecture beauty. Problems arise in this conceptualization when an engineer like Menn achieves both.

Such semantic schizophrenia afflicts more than the civil engineer. His colleagues in mechanical engineering, electrical engineering, aeronautical engineering, etc. do share common history with professions such as those of the sciences. Thermodynamics is required of most engineering students. Physics and mathematics are required at levels of proficiency where the distinction between engineer and scientist is not as easy to make as Professor Holbrow would have us believe. It is perhaps one of the less obvious reasons for this book. I would let the reader consider the development of engineering and draw his or her conclusions.

A far more serious reason for a text such as this stems from comments by other educators on the training of engineering students. I offer the following quote for the reader's consideration since it succinctly addresses the issue.

> The training of engineers tends to suppress rather than encourage a sense of history, but it is clear that many engineers transcend their academic background. As Donald Marlowe has pointed out the history an engineer learns gives him no sense of its ambiguity. ''The skimpy historical education of engineers,'' he writes, ''leaves them with a firm belief that (without qualification) George Washington was the father of his country; Eli Whitney was the father of mass production; Fulton did invent the steamboat.'' A little knowledge is, indeed, a very dangerous thing.[5]

[2] Saslly, V., Jr. *Louis Kahn*. G. Braziller (1962), p. 118.
[3] Guiedion, S. *Space, Time and Architecture*. 5th ed. Cambridge (1967).
[4] Le Corbusier. *Towards a New Architecture*, translated from the French *Vers une Architecture*. F. Etchells, London (1923), p. 16.
[5] Hartenberg, R. S. *National Historic Mechanical Engineering Landmarks*. ASME (1979), p. xxiii.

This is an introductory book. Therefore, more arcane, but important, issues of society and technology such as determinism and inter-connectivity are not addressed. The implications of engineered works and products for the environment are cogent topics but beyond the level of this survey. My purpose is to present persons, concepts and events that made salient contributions to engineering history. The impacts and implications of that history I leave to other commentators. My role is primarily that of reporter of a compelling story that spans millennia and my aim is to share with the reader my genuine admiration of engineering history.

DEDICATION

To My Children,
Ian and Brenna

THE AUTHOR

Ervan G. Garrison presently is Chief, Marine Archaeology and Maritime History Unit, National Oceanic and Atmospheric Administration (NOAA). Until 1990 he was a faculty member of the Environmental & Water Resources Division, Civil Engineering Department, Texas A&M University. During his eleven years at this institution he taught courses in the history of engineering and technology. This text is a result of this long interest in technical history and its role in society as a whole. He holds degrees in the natural sciences and anthropology, specializing in archaeology for his Ph.D. which he received from the University of Missouri in 1979. Over his career he has conducted research in physical dating techniques, ancient technology, and maritime archaeology. He has worked and taught in Europe and the U.S. He is the author of over 100 technical papers, journal articles, and contributions to books. Of Native American descent, he was born in Tulsa, Oklahoma April 11, 1943.

Acknowledgments

Chapter 1

Figure 1-4   Academic Press
Figure 1-7   Courtesy of Professor Joseph H. Senne

Chapter 2

Figure 2-5   G. Isfrah, *From One to Zero*. Viking (1985)
Figure 2-6   G. Isfrah, *From One to Zero*. Viking (1985)
Figure 2-7   G. Isfrah, *From One to Zero*. Viking (1985)
Figure 2-10   McGraw-Hill Inc.
Figure 2-12   Andromeda Oxford Ltd.

Chapter 3

Figure 3-1   McGraw-Hill Inc.
Figure 3-3   Courtesy of Professor O.A.W. Dilke
Figure 3-5   Harry N. Abrams Inc.
Figure 3-6   McGraw-Hill Inc.
Figure 3-8   University of California Press
Figure 3-9   University of California Press

Chapter 4

Figure 4-3   Council for British Archaeology
Figure 4-6   McGraw-Hill Inc.
Figure 4-8   Courtesy of Professor O.A.W. Dilke
Figure 4-9   Hammond Inc.
Figure 4-11   McGraw-Hill Inc.

Chapter 5

Figure 5-4
  (upper)   Courtesy of Lionel Casson
Figure 5-4
  (lower)   Courtesy of John F. Coates
Figure 5-5   Courtesy of Lionel Casson

Chapter 6

Figure 6-2   Courtesy of Professor Daniel MacGilvrey
Figure 6-3   Courtesy of Thorkild Schiola

Chapter 7

Figure 7-4   McGraw-Hill Inc.

Chapter 8

Figure 8-6   Courtesy of Professor Daniel MacGilvrey
Figure 8-7   Courtesy of Professor Daniel MacGilvrey
Figure 8-11   Burndy Library, Norwalk, Connecticut

Chapter 9

Figure 9-1   Penguin USA
Figure 9-3   Courtesy of Professor Daniel MacGilvrey
Figure 9-5   McGraw-Hill Inc.

The draft manuscript was prepared by Ms. Joanna Fritz.

TABLE OF CONTENTS

1 THE EARLIEST BUILDERS

Archaeology provides our earliest information on those works of man that could be considered as antecedents, or even artifacts, of the earliest engineers. During the long evolution of early man, technology, as a whole, is best characterized as conservative; innovations were singular and often separated by millennia or even eons in the earliest stages. The development of early stone tools is testament to this inertia in innovation. Surely this is, as anthropologists believe, the result of biological factors such as the evolution of the human brain. It is beyond the scope of this text to plumb these interesting issues in man's development other than to briefly recapitulate, in chronological form, the major advances in man's technological mastery leading to what we can recognize as engineering in practice and form.

The development of technological thought and, thus, the genesis of "engineered" works is directly tied to the evolutionary development of the human brain. We have no actual physical remains of ancient man's brain, but his development can be inferred from two related sources: (1) Pleistocene man's skeletal remains, and (2) Man's cultural remains. Table 1-1 plots the accepted general course of human and early technological development. We will begin our discussions with stone tool manufacture. Inspection of the chart shows the first appearance of identifiable tools at around 2.25 million years. B.P. (before present) These are shown in Figure 1-1.

The oldest pebble tools have been found in ancient (2–2.25 m.y.) sites in Africa along the margins of Pleistocene lakes or streams. The associated skeletal remains have been termed *Homo habilis* by Louis B. Leakey and were first identified at the Kenyan site of Oldavai Gorge along the Central African Rift Valley Chain. Campsites have been identified as associated with the remains of hominid tool users and are collections of rude shelters or windbreaks rather than structures.

Only by 500,000 years B.C. do we see campsites of extensive occupation with the demonstrated use of controlled fire. These occur in China at a place known as Chou Koutien, the type site for *Homo erectus sinanthropus* or "Peking Man". The tools typical of this phase of man's development are called core tools (Figure 1-1). Other members of *Homo erectus*, outside Asia, had a separate tradition of lithic technology in which core tools predominated. Combined with bone points and clubs, these stone tools allowed *Homo erectus* to occupy large areas of Africa, Europe and Asia. These groups were hunters of Pleistocene megafauna such as mammoth, giant sloth, camel, horse and bison.[1]

Excavations at Nice, France have yielded the first evidence of man-made habitation structures.[2] These shelters reached over 13 meters in length, were made of saplings joined along the ridgeline and upheld by four vertical supports. These structures were probably covered with leaves, although other materials such as skins might have been used. A perimeter of stones braced the lean-to walls on the windward sides. The analysis of pollen from human waste (coprolites) indicates a summer occupation of the site.

Throughout the Pleistocene epoch, spanning approximately three million years before the modern era (termed the Holocene), the engine of change was climate. The Pleistocene and, indeed, the whole Quaternary period, which includes the Pleistocene and Holocene, is that portion of the geological time scale dominated by glaciations which reshaped much of the northern hemisphere's land surface and concomitantly influenced man's development as a species.[3]

TABLE 1-1

	Years B.P.*
Farming and food production	10–12,000
Mechanical devices like bows and spear throwers	30,000
Representational art	30,000
Hafted tools appear (Mousterian)	50–100,000
Humans occupy the cold, temperate zone	300–500,000
Definite signs of controlled fire	300–500,000
Last definite signs of multiple species of coexistent hominids (Abbevillean/Acheulean)	1 million
Oldest definite "camp sites"	2 m
Oldest known definite stone artifacts (pebble tools)	2–2.25 m
Oldest fossil skull with a brain case much larger than that of any ape	2–2.5 m
Oldest known definite indications of a fully bipedal mode of locomotion	3–4 m

At this point there is a virtual gap in the fossil record between about 4 and 9 million years ago, and only a few uninformative scraps have been found.

Oldest known traces of hominoids with teeth and jaws like those of later hominids (at least three species: *Ramapithecus, Sivapithecus,* and *Giganthopithecus)*	9–14 m

*   Conversion to years B.C.: subtract 1950 years from B.P. date. B.C. will be more commonly used in this text.

The inhabitants of the simple camp at Nice were mid-Pleistocene (400–500,000 years ago) hunters living in a Europe which was enjoying a respite from the cold of the glacial ages. During the Pleistocene, at least four major glaciations have been verified, with numerous warmer or colder sub-stages of partial retreats or advances. These hunters and their simple stone axe technology ranged over three continents—Africa, Europe and Asia—during the middle Pleistocene until roughly the end of the third of the glaciations, termed the Riss.

About 100,000 year B.C., *Homo erectus* disappears in the fossil record and is replaced by more advanced forms such as *Homo sapiens neanderthalensis.* This early man used a much different stone technology called the Mousterian or "prepared core" tools. We are not sure how Neanderthal man related to a more modern sapiens form, but the differences are in degree rather than in kind.

As shown in Figure 1-1, Mousterian tools differ in appearance from other Paleolithic stone implements. For the most part, Mousterian artifacts are made of flakes, large in the main, struck from prepared cores. Neanderthal man lived during the early half of the last glaciation, the Würm, from 75,000–40,000 B.C. After a 10,000-year warm period, only anatomically-modern sapiens forms are seen. Associated with these men is a core-blade tool technology. In the Americas, men of this type are thought to have first invaded the continents during the onset of the last glacial phase, its concomitant lowering of sea level creating land bridges such as in Beringea between Asia and North America.

What is important to us in this discussion of early man is to form an appreciation of the long prehistory of man's slow technological development. Man's technology was simple and direct. We can speculate that this implies a direct relationship between man's needs and his intellectual capacity to find ways to meet these needs. We suspect this was the case, but can only infer this from our analysis of early man's fragmentary archaeological record. We do know that after 10,000 years B.C., we see engineering thought evident in an expanded technology that includes: the construction of structures and boats; construction of water reservoirs and irrigation canals; domestication and use of animals for motive power and the development of metallurgy. This "new stone" age is termed the Neolithic.

BLADE TOOLS

UPPER PALEOLITHIC

FLAKE TOOLS

MIDDLE PALEOLITHIC

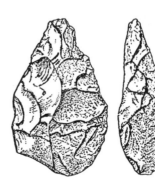

CORE TOOLS

PEBBLE TOOLS

LOWER PALEOLITHIC

FIGURE 1-1.   Early stone tools.

THE NEOLITHIC

The Neolithic marked the end of the Glacial Age, a hard and difficult time, with vast areas uninhabitable for plants, animals and man. Upper Paleolithic man did develop a rich way of life centered on hunting and gathering, but it was still a technology of stone and bone tools geared to a transient pursuit of game such as reindeer in the habitable reaches of Europe, Asia and Africa.

With the end of the glaciers, the environment warmed, spelling doom for Pleistocene animals such as the mammoth, cave bear, saber-tooth tiger, American horse and numerous other, now extinct, species. The lifeways built upon the pursuit of these animals ended as well. It is not the purpose of this book to detail the prehistory of man in all its richness. We are interested in man's engineering and its development. With the Neolithic, we are at its beginning, primarily because man had to build for permanency for the first time.

The development of primitive agriculture is one major definitive characteristic of the Neolithic Period. The origins of agriculture do not directly concern us, but the sedentary lifestyle farming engenders does have direct ramifications in terms of engineering and technology. In the early spread of Neolithic agriculture (11,000–6000 B.C.), farmers exploited loessic soils left by the glacial periods along major river valleys of Europe, such as the Danube. The early inhabitants of the post-glacial forests farmed these alluvial fans and loesic terraces of streams from the Mediterranean to the Baltic. Having thus implied a direct connection between sedentary life, agriculture and building one of the earliest known examples of planned, large-scale building comes, as an exception, from pre-agricultural levels of an early mid-eastern village, Jericho.[4,5]

Jericho is located near the Dead Sea in the Trans-Jordan (Figure 1-2). Here archaeologists have found layers of permanent settlement reaching 20 meters below the present ground level and representing over 7000 years of occupation. Why? Jericho is an oasis in a desert that lies 256 meters below sea level. A spring called Elisha's Fountain provided a copious flow to the nearby Jordan River, allowing men to raise by 8000 B.C., a settlement boasting a massive stone wall. This structure's base was a ditch, 8.2 meters wide and 2.1 meters deep, carved out of solid rock. Today, after 10 millenia, the wall is still almost 4 meters high. Inside the wall stood a circular stone tower (Figure 1-3), still preserved to a height of 8 meters.

The village this wall surrounded contained huts made of sun-dried bricks constructed into curved walls reminiscent of the tents or temporary shelters they surely replaced. This 4 hectare village raised early forms of wheat and kept goats. Aside from the formidable wall, the level of social organization is Neolithic, termed ''pre-ceramic'' owing to the lack of that friend of the archaeologist, pottery. However, as a pre-ceramic settlement, Jericho fits nicely into cultural evolutionary sequences that trace technology from hunter-gatherers of the late Pleistocene to the farming villages of the Neolithic. Ceramics have never been a necessary condition for cooking or food storage. Ground stone, wooden containers, baskets or leather work allowed many early cultures to manage nicely in this area. In the United States, we have evidence of Neolithic stage peoples in the California Indian groups who made wonderfully functional baskets up to the beginning of the 20th century. Pottery manufacture may have been one antecedent for man's next technological step—metallurgy. Still, the early residents of Jericho excavated their wall's footer trench without metal tools. This next technological step was to wait another three millennia before man smelted copper ores, into small ornaments initially rather than tools.

A Neolithic ''town'', near to and contemporary with Jericho was 'Ain Ghazal, which flourished from 9250–8000 B.P.[6] At its height, it had a population of 2500–3000 people. Ain Ghazal, lacking the fortification of Jericho existed beyond Jericho in time (5000 vs. 6000 B.C.). However, its last few hundred years show a marked decline in prosperity, building and population. Whatever caused Neolithic Jericho to be abandoned also affected 'Ain Ghazal, but the latter settlement changed or adapted to survive.

FIGURE 1-2.   Jericho is located in Jordan near the Dead Sea. It is in the midst of a desert, but has been made habitable for 7,000 years thanks to the waters of Elisha's Fountain.

West of Jericho, in what today is Turkey, developed perhaps the oldest city in history. Çatal Hüyük began around 8300 B.C. and flourished up to 5600 B.P. on the Anatolian Plateau as a town of some 6000 people, trading obsidian and raising wheat, barley and domesticated animals. To date, 139 buildings have been excavated representing only a small portion (0.5 hectare) of a 13-hectare settlement. Over a third of the rooms found were shrines to a cult or religion in which the bull was paramount. Bull images, painted and modeled in clay, protruded from walls and altars. The structures themselves were much like the pueblos of the American Southwest. They were of beam frame, mud-brick, flat-roofed with roof entry and upper windows. Other than sanctuaries, the houses contained storerooms and living areas. Their size averaged 6 by 5.5 meters.

In early-mid Neolithic Period Europe, we see nothing to compare to Jericho, 'Ain Ghazal or Çatal Hüyük. After the glacial retreat, Europe experienced cyclical periods of wetter or drier climate over hundreds of years. In general, this caused fluctuations in water levels of streams and lakes, the favored settlement areas of early Neolithic peoples in central Europe. The retreat of the ice caps was coupled with the advance of forests across central Europe. Where Paleolithic men had followed herds of reindeer across a taiga or steppe-like landscape,

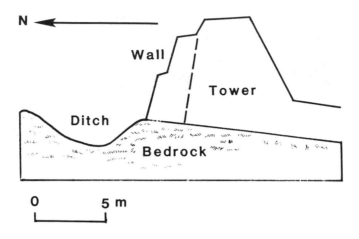

FIGURE 1-3.   Wall, ditch and tower of Jericho, ca. 6000 B.C. (Redrawn from Kenyon, 1960.)

men of the Neolithic had luxuriant, climax forests of mixed hardwoods to challenge settlers using a stone tool inventory. Because of the forests, the Neolithic peoples of what is now western France, Switzerland, northern Italy, Germany and Poland utilized wood as their principal building material.[7]

By 7000 B.P., they built large houses, "longhouses" if you will, that reached up to 30 meters in length.[8] From the Danube to the Netherlands, these structures demonstrated an almost standardized form of building. With widths of 5 to 8 meters, these rectangular buildings were built on five rows of posts of which the inner three upheld the roof, while the outer two rows supported walls made of interwoven branches packed with clay. The surfaces of the walls were smoothed and whitewashed on the interior. Oriented northwest-southeast or north-south, the northern end was reinforced with posts sunk in trenches and spaced closely (1.5 meters as compared to up to 3.5 meters). The houses were divided into three sections—a central living area, a southern storage area and a northern stable for livestock.

Gradually, from 2600 B.C. to 1000 B.C., European cultures began the monumental use of large rocks or "megaliths", some weighing up to 30 tons.[9] The forms of construction included grave chambers, upright stones (menhirs) often aligned in circles or avenues, barrows and tumuli. Originally, these megaliths were thought to result from Mediterranean and ancient Near Eastern cultural influences. Today, the evidence is compelling for these structures to predate those of the Mediterranean rim.[10] The information gained through the use of modern physical dating techniques suggests that Europeans of the fourth to second millennia had developed a level of civilization, including metallurgy, quite independently of the Mediterranean cultures. In other words, architects or engineers from Grecian Mycenae did not build monuments such as Stonehenge on the Salisbury Plain of western England.

STONEHENGE

In the late Neolithic and early Chalcolithic ("Copper Age"), Periods a simple earthen structure was constructed on Salisbury Plain. The original construction consisted of a circular ditch with a diameter of 9 meters (30 feet). The depth of the ditch was 1.5 meters (5 feet). The embankment enclosed 56 shallow pits, some containing cremations. Radiocarbon dates for some of these are 2940 ± 150 B.C. The earthworks are roughly oriented toward sunrise at the summer solstice. A banked road approached from the northeast.

In the 16th century B.C., the entire circular site, which had a diameter of 30.7 meters, was enclosed with 30 upright sarsen stones,[11] each rising 4 meters above the ground and weighing 26 tons (Figure 1-4). The uprights had lintels cut to the curve and joined by tongue-

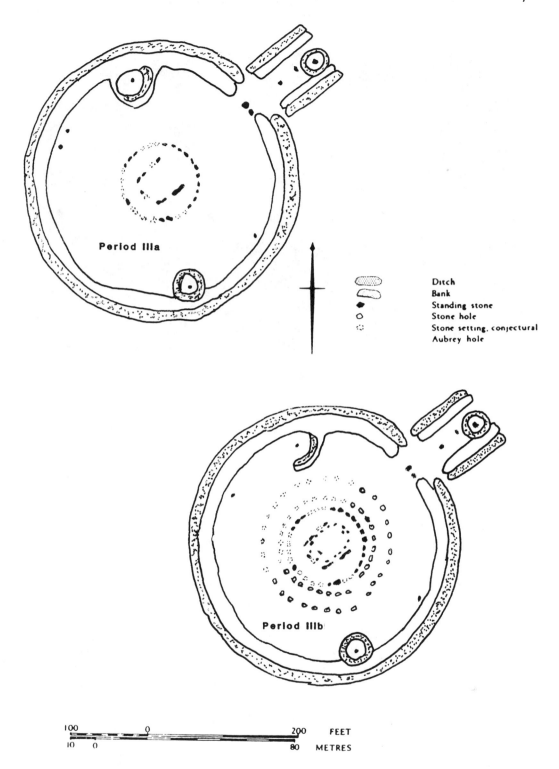

Period IIIa

Period IIIb

Ditch
Bank
Standing stone
Stone hole
Stone setting, conjectural
Aubrey hole

100   0   200   FEET
10   0   80   METRES

FIGURE 1-4.   Stonehenge, plan views during the early Bronze Age, ca. 1500 B.C. (Redrawn from Champion et al., 1984.)

FIGURE 1-5.   Stonehenge, view through sarsen ring to heelstone. (Photograph by author.)

and-groove. At each end, the down side of the lintel had a mortise hole that fits a corresponding tenon on the top of the uprights. Each upright had two tenons (Figure 1-5). The 20 lintels are secured into a circle of stone. The rock was quarried at Avebury in Wiltshire, 30 kilometers from the site, and hauled overland, roughly 50,000 ton-kilometers. The stones were dressed on-site. The outer sarsen ring encloses a concentric ring of "bluestones" that taper to a height of 3.2 meters. This bluish-grey sandstone is found only in the Presselly Mountains in Pembrokeshire, 300 kilometers from the site.

Inside the bluestone ring are five so-called "trilithons", placed in a horseshoe arrangement. Each lateral trilithon consists of two 6-meter sarsen uprights carrying a tenoned lintel, while the center extends 6.7 meters above ground and weighs 40 tons (Figure 1-6). Inside the trilithons is an ovoid of bluestones surrounding the "alterstone". The orientation of the monument has the center trilithon and "alter" slab in front of it facing the rising sun at the summer solstice. The slabs were hammered with stone mauls. The uprights were given entasis, or corrected, for the illusion of concavity by being cut and finished slightly curved.

The bluestones, quarried in Pembrokeshire, very probably were transported by barge down river to Milford Haven and thence along the coast to the mouth of the Avon and up the Avon to its tributary, the Fome. The stones would have to be transported over the divide to the river Wyle and hence to the site. The work amounted to 500 tons transported 200 kilometers, or 10,000 ton-kilometers.

Clearly, the work was carried out over many decades, probably during agricultural off-seasons. It was directed by an authority of continuity and purpose. What this authority, or its ultimate purpose, was is purely conjectural. R.J.C. Atkinson summarizes the complex history of Stonehenge as follows: *Stonehenge I*. Around 2000 B.C. (4000 B.P.) (or 2500 B.C. by recalibrated dates)—construction of the ditch and ritual shafts or pits. Erection of the "heel stone" for alignment. *Stonehenge II*. Around 2100–1700 B.C. (4100–3700 B.P.)—building of a double circle of the bluestone ring. The quarries were in Pembrokeshire, South Wales, 300 kilometers distant. *Stonehenge III*. This phase has been assigned to the early Bronze Age Wessex Culture. Three subdivisions, A, B, and C, correspond to modi-

FIGURE 1-6.   Stonehenge, trilithon. (Photograph by author.)

fications in the "bluestone" structures, but the major efforts were in the erection of the sarsen ring and the trilithons. Precision techniques were needed to quarry, transport them 30 kilometers and raise the stones. By 1100 B.C. (3100 B.P.) certainly, construction at Stonehenge was complete.[12]

Since Stonehenge was built in a series of phases, we cannot point to some single engineer of genius for its inception. We can, however, appreciate the talents of several men who designed and built this structure without the crush of technical and administrative paperwork so common to today's builders. They assuredly understood the geometry of the circle and right triangle. This knowledge is hinted at in recurring commonalties in measurements from several other rings, including Stonehenge.

During the last century, a scholar by the name of René Kerviler worked out a system of megalithic measurements. His basic unit was the megalithic foot (0.3 meter). The megalithic pace was three English feet and a megalithic chain was equal to 30 paces. René Merlet, a student of stone circles, refined the megalithic foot to 0.3175 meter. In our time, Professor Alexander Thom has examined data from 112 well-surveyed English and Welsh circles whose mean diameters are known to within 0.3 meter. From his study, he has derived a value for the megalithic yard of 0.829 meter or 2.72 feet. If these hypotheses are correct, then the builders of Stonehenge used a basic unit of 30 centimeters and treble multiples thereof.

Simply from the standpoint of engineering, Stonehenge is as fascinating to us as Neolithic Jericho's fortifications. Stonehenge in plan (cf. Figure 1-4) gives us a close approximation of its final form, although the ages have left it the impressive ruin of today. A partial reconstruction at one-half scale was accomplished in 1984 by faculty and students of engineering at the University of Missouri-Rolla. This replica allows us a view of how the original may have appeared on a Bronze Age solstice (Figure 1-7). The replica represents the structure during the Stonehenge III Period (1700–1100 B.C.). The trilithons dominate the reconstruction as the Missouri engineers have reduced the sarsen ring to 29 marker stones in place of the circular arcade of the original.

In the sarsen ring, the first Stonehenge builders created a circular structure where the sum of the vertical forces acting on the beams is in equilibrium or is zero. Further, none of the beams are in shear or have stress forces acting on them to any significant degree. With the trilithons, the lateral support of the adjacent lintels have been foregone to create a construction which contrasts massiveness with free-standing form. This is best seen in the replica by the intentional reduction of sarsen ring.

Much debate exists over the function of Stonehenge or other megalithic "henges." In recent years, it has been suggested as serving an astronomical function, as a ritual or cult sanctuary, or as an ancient administrative or political place such as that of Iceland's comparatively younger (ca. 900 A.D.) Althing. Based on extensive study of over 900 British stone circles, one frequent feature is a ditch or bank enclosing the area. The ditch is *inside* the bank, such as at Stonehenge's larger neighbor, Avebury, which signifies a non-military function. Further, the presence of a high bank, such as needed for fortification, militates against astronomical observations as well. Two further distinctions occur; one type of henge has a single entrance through the bank and ditch, while the second type has more than one. What these differences mean remains to be discovered.

Our understanding of these early builders is hampered, in part, by the fragmentary record with which we are left after such a long passage of time. This is the nature of prehistoric study; only that which is durable enough to withstand the ages, such as stone, or which has been fortuitously preserved remains for the modern observer to interpret. A knottier problem in our understanding of these past societies and their technologies has been addressed by Steven E. Falconer of New York University.[13] The issue is our ability to interpret a society that has no modern analogue and is, therefore, outside our experience. Falconer suggests that many early cultures may represent extinct forms of society. The society that built Stonehenge may represent such a case; we cannot know their purpose or methods with any exactness. We are left to admire, and even marvel at, their works with an educated, yet profound, ignorance. The past is Louis Carroll's "looking glass" which allows us to only glimpse darkly.

Homo erectus and Neanderthal man were hunters such as the world will never see again. We can only imagine the world they knew. For that springtime, 350,000 years ago, the camp at Nice required early man or man-like beings to use cooperation in their construction of some of the first "planned" structures. Neanderthal man survived environmental conditions that would pull down our present-day specialized technology unless we displayed a similar adaptability.

A puzzle remains as to the origin of plant and animal domestication. We have the general picture with the early hillfolk of Iran and Iraq who tended descendants of wild grasses and game. This allowed these groups to settle into small villages. Rather than following their game or being compelled to move from harvests of nut mast to seed crop, men could carry their grains in their pouches. They did this, following the rivers and their alluvia across the

FIGURE 1-7. Solstice at Stonehenge replica. (Photograph courtesy of Professor Joseph H. Senne.)

Levant, the Balkans and into Europe. In China and the New World, other men did likewise. In each of these places, man adapted his technology to the local environment. Engineers appeared as designers and builders of irrigation systems, fortifications, public buildings and great tombs. These men and their works arose from a prehistoric horizon more quickly as agricultural techniques caused a quantum leap in man's population.

Hamblin has said that agriculture is not deterministic of cities or urbanization, but civilization is synonymous with cities. Populations grew rapidly in the fourth millenium, such that the children of the early farmers had to search for new lands to cultivate. Their neolithic agricultural techniques accompanied them. It was mentioned at the beginning of this chapter that innovation at first was limited by physiological factors, e.g., the development of the human brain. DeCamp, in his history of ancient engineering, has suggested a link to population as well.

If we grant early men the ability to reason, and certainly by Neanderthal times the brain size, if not structure, was modern, why the glacial slowness in innovation? DeCamp argues that primitive societies are conservative, owing in great part to their lack of economic surplus.[14] Experiments, particularly failed ones, in the area of a subsistence technology could threaten these societies with death. The old adage, "if it works, don't change it," may have a greater antiquity than we suppose.

Still, these primitive technologies could not support large populations. Men lived in small bands, probably organized through extended family ties. The areas needed to support such groups were large, based on a prehistoric carrying capacity of a predator-prey balance. As good as these hunters were, they were still inefficient as killers and wasteful of their resources to a great degree. Only when the forests cut off the migration routes of the great herds was man forced to consider alternate strategies to survive. When man "considered" the edible equivalent of the field lily, only then did he begin the trek along the path to agriculture, economic surplus, population and civilization. This trek led early hill-farmers of the Zagros mountains to chance the plains of the great Tigris-Euphrates valley with their more fertile soils. Here innovation took wing as villages became towns and towns became cities.

In Mesopotamia, as this area became known, development began in the middle of the fourth millenium with the Ubaid culture in the lower, or southern, downriver part, calling themselves Sumerians by 3500 B.C. This period saw the development of writing and metallurgy, together with building on a scale unseen heretofore. Evidence of this genesis is seen in the centralization of the population of Uruk.

From 146 villages in 3000 B.C., to 76 villages 300 years later, and a final 24 in 2400 B.C., the villages were not depopulated by disease or war.[15] The population moved to cities such as Uruk, Ur and other early city-states or poli, as the later Greeks would call them. It is here we begin our inquiry into the conquest of materials by the engineers of these young civilizations.

NOTES

1. **Butzer, K.** *Environment and Archaeology.* Aldine (1971), p. 448.
2. **de Lumley, H.** A paleolithic camp at Nice. *Sci. Am.* **220**: 42–50 (1969).
3. **Attenborough, D.** *The First Eden.* Little-Brown (1987).
4. **Hamblin, D. J.,** *The First Cities.* Time-Life (1973).
5. **Kenyon, K. M.** Ancient Jericho. *Hunters, Farmers, and Civilizations: Old World Archaeology.* Freeman (1979), pp. 117–122.
6. **Rollefson, G. and Simmons, A. H.** The life and death of 'Ain Ghazal. *Archaeology* **40**(6): 38–45 (1987).
7. **Osterwalder, C.** *La Suisse Préhistorique.* Editions 24 heures (1980).

8. **Schutz, H.** *The Prehistory of Germanic Europe.* Yale (1983).
9. **Burl, A.** Dating the British stone circles. *Am. Sci.* **61**: 167–173 (1973).
10. **Renfrew, C.** Recalibrated radiocarbon dates place the final construction date at 1800 B.C. *The Emergence of Civilization.* Alfred A. Knopf (1975).
11. Sarsen = hard sandstone.
12. **Atkinson, R. S. C.** Moonshine on Stonehenge. *Antiquity.* **43**: 212–216 (1969).
13. **Falconer, S. E.** An early 'Dark Age' in the southern Levant—insights on the basis of complex society. Paper presented at the 1st Joint Archaeological Congress, Baltimore, MD (1989).
14. **DeCamp, L. S.** *The Ancient Engineers.* Ballantine (1963).
15. **McEvedy, C.** *The Penguin Atlas of Ancient History.* Penguin (1967).

2 EARLY EMPIRES AND THE CONQUEST OF MATERIALS

As we have seen in the preceding chapter, man, in the Paleolithic or "Old Stone Age", confined his use of materials to the manufacture of tools. Man, in the Neolithic Period, continued the use of stone for tools such as hoes, axes, weapons and hunting gear. The Natufians, who probably founded Jericho, hunted gazelle, exploited wild cereal grasses, as evidenced by stone sickle blades, and lived in 0.25 hectare settlements. By 8500 B.C. (10,500 B.P.), Jericho was a pre-farming settlement. Later, farming was added along with extensive walls and towers. Except for rare exceptions such as Skara Brae, a neolithic farming outpost in the storm-swept Orkney Isles off Scotland, and Jericho, the early farmers normally did not build with stone.

One reason for this lack of stone construction was the type of polity inherent to this level of human society. Men lived in small villages of interrelated communal groups. This was reflected in their structures, common long-houses with living space for more than one family. In the forested regions, the people utilized wood for construction; in Egypt, stone; in Mesopotamia, brick; and in China, wood.

In five principal areas across the world, plant and animal domestication preceded apace. These "nuclear" areas were:

1. The Tigris-Euphrates Valley
2. Egypt
3. China
4. Southeast Asia
5. Mexico and Peru

It is no coincidence that these areas were the first to develop state-level societies and engineered works. An exception to early wood construction was in Mesopotamia, where people of the Tigris-Euphrates Valley-Plain, without such resources, used mud and reeds. It was in Sumeria that we see the translation of one structural material into another form of that same material—mud to brick.

The early Sumerian societies of the Tigris-Euphrates delta built houses of reeds that grew to heights of 10 meters (30 feet), lashed together into tall, thick bundles called *fascines* that were placed as uprights, equidistant along a straight line. To these vertical members were lashed thinner fascines laid horizontally, whereby they formed the framework for a wall. To complete the wall, the spaces between the framework were filled with reed matting. The two gables were constructed similarly. The final element was the roof—tall, vertical fascines were bent inward and lashed together where they met. Reeds were tied to these rafters with matting, thus completing the roof. The roof could assume, depending upon the extent of bending, either a semicircular vault or a pointed arch (Figure 2-1).

The design of an early structure has been described, but how does a reed construction lead to the origin of brick? The Sumerian reed house gave shelter until wind or rain came; the next step was to seal the drafty walls and leaky roof with mud plaster. This left only one serious defect in the design—it was a fire trap. An ember (or enemy) could ignite the dry reeds, whereupon the house burst into flame instantly. Where the house was covered in plastered mud, the burnt shell often stood after the fire that consumed the frame. Man, in all likelihood, developed the concept of the masonry structure from such chance experiences.

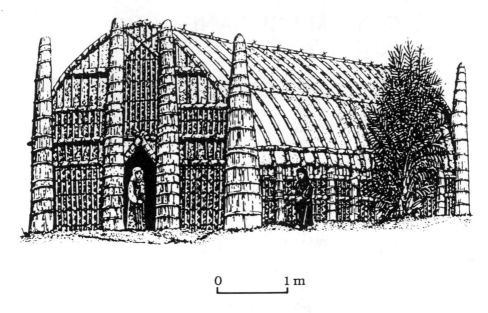

0 |_____| 1 m

FIGURE 2-1.   Reed-fascine house of the 4th millenium Mesopotamia. (After Leacroft, in Lloyd, S., *The Archaeology of Mesopotamia*, 1978.)

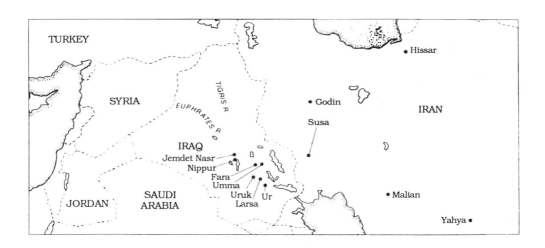

FIGURE 2-2.   Early Sumeria ca. 2000 B.C.

EARLY SUMERIAN SOCIETY

The first significant change in human life to follow the development of food production was the rise of cities, which occurred sometime before 3000 B.C. (5000 B.P.). The principal difference between a village and a city was that most of the inhabitants in the village were directly engaged in the production of food, whereas very few city dwellers were so engaged. Imposing cities and buildings did not arise overnight. For example, at Hassuna (Northern Mesopotamia) we see the transition of building techniques.

At Hassuna, the evolution of architecture from house to temples and in Southern Mesopotamia, the evolution of tools from bone and stone to metal occurred. Writing appeared at Uruk (Figure 2-2) by 3000 B.C. Their mathematics was based on 60, or sexagesimal, which we still use today for units of time (hours, minutes, seconds) and divisions of the circle. A factor of 60 divides our year into months and feet into inches.

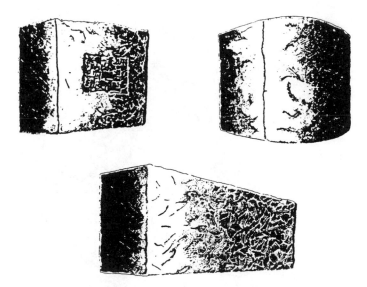

FIGURE 2-3. A variety of bricks was used by Mesopotamian builders at different times. The most common was a rectangular shape (bottom) made of mud with a straw binder. The babylonians also molded square units, often bearing inscriptions to their gods (top left), and the Sumerians used loaf-shaped bricks (top right).

It should be noted, however, that most early knowledge was purely empirical, gained from experience and handed down from generation to generation. Computational techniques and arithmetic gradually came to be used for commercial purposes and in surveying and mensuration in third millennium Sumer. All of these changes had their influence on the engineering of the period. The growth of the cities stimulated engineering in other ways. Before 3000 B.C., most buildings were reed or mud-clay homes. After that time, structural engineering was no longer merely functional, but architectural as well.

The most important sources of information about the engineering of Mesopotamia are the ruins of the cities themselves. Indeed, the work of these archaeological investigators has provided much of our understanding of these ancient sites. Our perception has been colored, however, by the manner, or order, in which the ruins have been excavated. Certainly, the precedence in the historic research does not imply any chronological order or cultural importance of one site over another. Our knowledge continues to grow today with modern discoveries such as the library of Ebla. Archaeologists first found at Ur, south of modern Baghdad, Iraq, enough of the ziggurat, or temple tower, to give us a very good idea of what it was like in its original form. High walls cannot be made with sun-dried brick. Masonry of this type cannot withstand much heavy pressure or compression before it fails. To relieve this vertical pressure or static load, they started a second wall behind and above the first, each upon an interior mound of earth. They often built three or more walls. The outline of the taller ziggurats must have been quite similar to our modern skyscrapers with set-back stories.

The remains at Ur indicate that its ziggurat was a many-angled and solid pyramid, approximately 70 meters (200 feet) by 40 meters (150 feet) and 20 meters (70 feet) tall. It was made of sun-dried brick, faced with burnt brick and some stone (Figure 2-3). There were a number of stairways by which people could ascend to open-air shrines at the summit. The material commonly used to bind the masonry was asphalt or bitumen; it still clings to bricks laid 25 centuries ago. The prize of international rivalries, it is found, to this day, bubbling from the ground near the present-day site of Baghdad (Iraq).

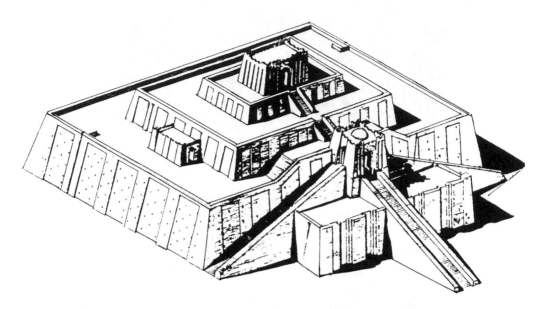

FIGURE 2-4.   The Third Dynasty ziggurat of Ur-Nammu at Ur. The overall dimension at ground level is 61 × 45.7 m. (Lloyd, 1978.)

The builders further relieved pressure on the brick facade and drained water by the use of reed matting mixed with the bitumen. Every few tiers, they would use this structural device as they ascended to the final height of the construction (see Figure 2-4).

The cities also produced problems which the evolution of hydraulic engineering gradually solved. Open drains were constructed to remove surface water, and in a few instances, underground drains kept building foundations from sinking into mud. The need for increased efficiency in food production resulted in the construction of levees, dams, reservoirs and canals for flood control and irrigation. City dwellers needed water, and tunnels were sometimes constructed to bring water from nearby springs to pools within the city. A deciphered clay tablet, dating from around 1200 B.C., refers to a form of treadmill for raising water from an irrigation ditch. The tablet is a legal text or document acknowledging that a man had borrowed a water-raising wheel of 17 steps or 6 meters; it had been lost and he must replace it.

The mounds and ridges which remain show that the embankments and levees protecting the lands of the lower Euphrates were 30 meters (100 feet) wide and hundreds of kilometers long. Wherever feasible, there evidently were spillways to carry excessively high water into great depressions in the desert. These depressions covered as much as 1,700 square kilometers (650 square miles), and they could hold water to a depth of 7 meters (25 feet). Such storage capacity may have been necessary given the net waste of irrigation even today. Irrigation requires a large amount of water. In distribution, about 50% of the water gets lost and 25% flows away unused, leaving 25% to be used on the crops.

Sumer's, and all of Mesopotamia's, sources of water were the Tigris and Euphrates. The two rivers are quite different in their volume and flow. Rising earlier than the Euphrates, the Tigris carries about 2.5 times as much water, but it is more incised, or deeper, than its partner. Even though the Euphrates carries only 40% of the volume of the Tigris, it provided the bulk of irrigation waters since its bed was raised above the surrounding plain. This geomorphology allowed gravity flow of waters into the canals. The Tigris was used to water lands on its left or west bank. Rising in the wrong season, the rivers forced the Mesopotamians to practice perennial irrigation, where storage was mandatory. Table 2-1 gives a comparison of Mesopotamian irrigation conditions with those of its neighbor in antiquity, Egypt.

TABLE 2-1
Comparison of the Conditions of Irrigation in Ancient Egypt and Mesopotamia

	Egypt	Mesopotamia
Season of the floods	August to early October	April to early June
Climate	Semi-tropical	Continental
Average summer temperature	43°C (110°F)	48°C (120°F)
Average winter temperature	12°C (53°F)	6°C (40°F)
Season after the floods	Winter	Hot summer
Relation of harvest and floods	In time for winter and summer crops	Too late for winter crops Too early for summer crops
Rise and fall of the waters	Slow and clear rise and fall	Sudden rise and fall
Profile river valley	Concave, sloping towards sea No stagnant water	Very flat, faintly sloping towards sea Pools and swamps
Surrounding country	Lime and sandstone hills	Weathered marls containing salt and gypsum
Type and quantity of sediment	Sufficient, salt-free sediment; little silting-up of canals	5 times as much salty sediment, canals silting-up quickly
System of irrigation	Basin irrigation	Perennial irrigation
Effects of irrigation	As result of irrigation and type of soil tendency to extract the salts present in the soil; very slow silting-up of canals	Tendency of salts and alkaline compounds to accumulate in soil; danger of silting-up of the many canals

The early builders of Sumer scored some of their most significant achievements in hydraulic engineering. The interaction of hydraulic control and the bureaucratic state has been commented on by several authors. Most notable of these were Karl Wittfogel[1] and Julian Steward,[2] who pointed out the simultaneous development of early civilization and large-scale agricultural irrigation. Wittfogel's hypothesis stresses the importance of water as a resource which required centralized control and coordination. Wittfogel described the range of activities necessary in a society that relies on large-scale irrigation: planning and construction, scheduling, maintenance and defense of the irrigation infrastructure of canals, reservoirs and control structures (locks and dams).

While reasonable and logically appealing, Wittfogel's ideas have suffered from two major criticisms.[3] The first objection arises from the archaeological and historical observations that large-scale irrigation was not prevalent in Mesopotamia until long after the rise of the state. Secondly, studies of modern Mesopotamian societies have shown that large-scale works are not necessary for an adequate agricultural livelihood. While damaging, these criticisms do not deny that managed irrigation and the state are not somehow intertwined.

Mitchell[4] has recast Wittfogel's ideas in such a manner that central coordination resulted in greater political integration, and this organization was expanded because of its economic advantages. In this model, the central government or bureaucracy grew incrementally with the irrigation infrastructure. In this way, a feedback relationship would have existed between increasing irrigation and growing authority of the government. The role of management was important to, but not wholly deterministic of, the urban state.

Urban development, with its political, economic, religious and social structures, stimulated and interacted with the development of engineering. In turn, engineering influenced politics, economics, religion and social life by making available the means by which these activities could continue to evolve. Engineering came into being to solve problems of these new societies and once established, there arose an interaction by which engineering, in turn, influenced the evolution of society.

MATHEMATICS AND WRITING

Sumerian engineers understood right triangles and computed areas of land, volumes of

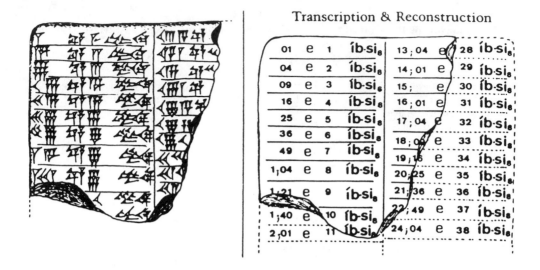

FIGURE 2-5.   Fragment of a table of square roots, with the numbers expressed in the Babylonian place-value notation. It dates from about 1800 B.C. and was found in the ruins of Nippur (about 100 miles southeast of Babylon). (Redrawn from Ifrah, 1985.)

masonry and the cubic contents of a canal. The evolution of numbers and measures must have followed very closely the beginnings of literacy in Mesopotamia. Friberg,[5] in summarizing his work and that of others, finds that by the end of the fourth millennium B.C. proto-Sumerian scribes had well-developed systems of numbers and measures, together with a precursor to our decimal system. Beginning with the tally, these early cultures used clay tokens continually from the ninth millennium B.C. to designate numbers, measures and categories of objects. Later, the token forms are seen as shapes on clay tablets.

These writing tablets, like the bricks of their structures, made use of one of southern Mesopotamia's few abundant resources—clay. It was better to impress clay than to incise it. The Sumerians began to impress their signs, using styli that were triangular in cross-section and giving rise to their wedge-shaped (cuneiform) writing.[6] First described by Sir Henry Rawlinson in 1855, these Sumerian tables of roots were found at Larsa, a neighboring city to Uruk (see Figure 2-5). Rawlinson determined that these texts, called Old Babylonian, represented a sexagesimal (base 60) number notation with a quasi-positional nature wherein the symbol for 1 also stood for powers of 60.

Sumerian writing originated about 3200-3100 B.C. and that of Proto-Elamite about 3000 B.C. Comparisons of the two rule out a common origin even with their close geographical proximity. Both seem to have been invented for strictly utilitarian reasons such as keeping up with economic transactions. The writing is termed ideographic in that it renders concepts such as "walk", "take", "plow", etc. into symbols. It is a pictorial writing system based on common signs for these concepts. It required a large number of signs, about 2,000, and suffered from ambiguity.[6]

The numeral system was less cumbersome, requiring an economy of only six signs. These are shown as follows:

1   10   60   600   3600   36,000

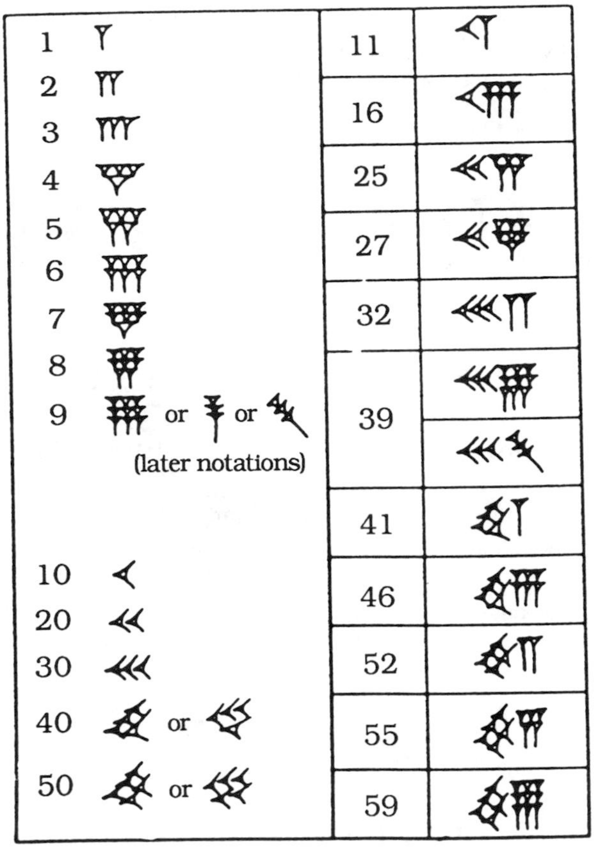

1	Y	11	◁Y
2	YY	16	◁YYY
3	YYY	25	◁◁Y
4	▽	27	◁◁▽
5	▽▽	32	◁◁◁YY
6	YYY	39	◁◁◁YYY / ◁◁◁
7	▽	41	⬦Y
8	▽	46	⬦YYY
9	YYY or ⅀ or (later notations)	52	⬦YY
10	◁	55	⬦YY
20	◁◁	59	⬦YYY
30	◁◁◁		
40	⬦ or ⬦		
50	⬦ or ⬦		

FIGURE 2-6. Representation of the 59 units of the first sexagesimal order in the Babylonian place-value system. (Redrawn from Ifrah, 1985.)

By the 27th or 26th century B.C., the non-circular signs underwent a 90° counterclockwise rotation:

⊃	∘	▷	▷∘	◯	◎
1	10	60	600	3600	36,000

Unless one is familiar with ancient Sumerian notation, the tendency is to refer to this system as "cuneiform", a term many of us know. In fact, it is not cuneiform or "wedge-shaped". This development followed shortly on the heels of the sign rotation of the 27th–26th centuries. This change resulted from the replacement of a round stylus or pen by a wedge-shaped one. The new signs became:

The use of these numbers is illustrated in Figure 2-6.

The later system, as shown in the Babylonian tablets, used the notion of place-value. By so doing, the Babylonians reduced their numeral signs from 59 to 9. A comparison between the earlier Sumerian system and the later Babylonian is seen in the writing of the same number, 69.

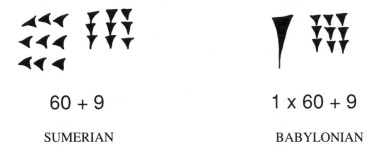

60 + 9	1 x 60 + 9
SUMERIAN	BABYLONIAN

This place-value notation originated with Babylonian mathematicians and astronomers. It became the standard for calculation in ancient Mesopotamia and may have influenced later Egyptian systems.

Some confusion was engendered in this system unless the user was careful to eliminate ambiguities such as:

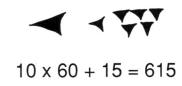

10 x 60 + 15 = 615

which could be confused with the notation for 25:

25

To avoid this, the users separated the symbols with blank spaces or utilized a sign to indicate a blank. These blank signs came to represent the first genuine zero used in mathematics.

Why the first number systems were based on 60 rather than on 10 such as ours is today is lost to memory. Friberg points out that it was a good choice as our decimal system has the weakness that its base of 10 is too small. Further, the Sumerians and their successors constructed a nicely interrelated measure system of standard units with conversion factors such as 2, 3, 5, 6 and 10 as compared to our conversion from the inch to the mile using factors such as 12, 3, 5¼, 4, 10 and 8.

Further aspects of Mesopotamian civilization include development of wheeled transport, legal codes and pyrotechnology (manufacture of glass, metals). The chronology of the Sumerian area is shown in Table 2-2 along with that of the Nile Valley, which we shall now examine.

EGYPT — MASTERS OF STONE

The rise of Egyptian civilization can be traced to before 4000 B.C. The state was unified after long periods of internal struggle under Menes, also called Narmer, in about 2950 B.C.

TABLE 2-2

Chronology — Sumerian Area

534–330 B.C.	Persian Empire
612–534 B.C.	2nd Babylonian Empire
1350–612 B.C.	Assyrian Empire
1990–1790 B.C.	1st Babylonian Empire
2900–1990 B.C.	Sumerian Civilization

Chronology — Nile Valley

1085–30 B.C.	Late Period
1567–1085 B.C.	New Kingdom
1786–1567 B.C.	2nd Intermediate Period
1991–1786 B.C.	Middle Kingdom
2135–1991 B.C.	1st Intermediate Period
2658–2135 B.C.	Old Kingdom
3100–2658 B.C.	Archaic Period

Egypt passed slowly from kingdom to empire. One significant stage early on in that process was marked by the pyramids which were built in the Old Kingdom period from approximately 2658 to 2135 B.C. Lasting over 500 years, the Old Kingdom was a period of prosperity, stability and confidence wherein Egyptians indulged their ambitions to a degree that matched or even surpassed the Mesopotamia of the same era. With the fall of the Eighth Dynasty, Egypt passed into a period, the First Intermediate (2135–1991 B.C.), of political disunity. The Middle Kingdom began with the conquest of the Heracleopolitans (ca. 2050 B.C.) by Thebean princes of the Eleventh Dynasty. The capitol was transferred to Memphis and the civilization of Egypt flourished anew. The southern frontier reached the Nile's second cataract with fortresses such as Buhen (ca. 2000 B.C.) being constructed. The Pharaohs began to challenge an expanding Hittite Empire based in Anatolia (modern Turkey), in the area of Palestine.

The Middle Kingdom ended in a collapse of unity between Upper and Lower (southern and northern) Egypt. Almost simultaneously, Egypt suffered its first real foreign invasion by the Hykos, or "sea peoples", which is represented by the Second Intermediate Period of the Fifteenth to Seventeenth Dynasties. Again, the Theban princes reestablished a unified Egyptian rule and by 1567 B.C. had founded the Eighteenth Dynasty. This is the period of Imperial Egypt with rulers such as Seti I, Ramses II, Akhenaton and Tutankhamon. During the Late Period Egypt slowly slipped from preeminence, being conquered by the Assyrians, Persians and, finally, the Romans. One of the last of the pharaohs of old was Shoshenq I who, in 925 B.C., conquered the Jerusalem of Solomon's inept sons, destroying or carrying off the treasures of Solomon's Temple.

Egypt, at its height, was one of the most populated countries in history. A population of eight million inhabited an area of roughly 29,000 square kilometers. Protected on the east and west by desert and on the north by the sea, the Valley of the Nile stretched southward into Africa, providing arable land to support this high population density (275 persons per kilometer).[7]

IRRIGATION

Irrigation was fundamental to the survival of Egypt. As in Mesopotamia, nature could not be relied upon to always provide a steady and adequate supply of water. The Nile provided the water for basin irrigation, which the Egyptians managed as a science.

One of the most notable reclamation projects in history was planned and carried through to completion by the Theban Dynasty of 2000 to 1788 B.C. This dynasty had reunited Egypt after the old regime of pyramid builders crumbled and turned an empire into the warring feudal states of the First Intermediate Period. The Theban kings developed an active trade with other parts of the Mediterranean and extended their influence beyond the borders of Egypt as far as Palestine. For the benefit of their people at home, they undertook to change the great oval basin in the Faiyum desert. West of the Lower Nile, this area became a fertile and populated land. They threw dams across the ravines leading into the basin to impound the rains of the wet season against drought. One of these dams in a 250-meter wide gorge had a base 143 meters thick—four times its height. It was built in layers, the bottom of rough stones embedded in clay, the next of irregular limestone blocks and the top of cut stone laid in steps so that the water pouring over the brim was checked in its fall and would not erode the structure. Egypt could boast a total of 884,600 hectares of arable land in the Valley proper; 1,402,600 in the Delta, and 123,000 in the Faiyum.

Egyptians in the third millennium B.C. were using engineering methods conditioned by their environment. They were still using those methods 30 or more centuries later when their country had ceased to exist as a separate entity. Three basic factors determined the nature of Egyptian engineering. One was the great supply of human labor. The horse was never used for work and was unknown in Egypt until about 1700 B.C. It is a mistake to think of Egyptian laborers as always cringing under the lash of an overseer, notwithstanding the biblical story of Moses. The Egyptian peasant fully expected to be drafted for public tasks in that portion of the year when the climate and the river kept him from his own work in the fields. He seems to have worked rather willingly, took pride in the enterprise and was amenable to the discipline necessary to govern hordes of men straining at the same task. Inscriptions boast that work teams accomplished tasks without accident or sickness and on time. Another determining factor in Egyptian engineering was the concentration of these vast armies of workmen under the absolute control of a single man and his lieutenants. Egyptian engineering was that of the technical overseer and its achievements were on a scale befitting pharaohs who called themselves both king and god. A third factor in Egyptian engineering was the great quantity of building stone in the ledges of the upper Nile Valley. Unlike Mesopotamia, Egypt was almost completely self-sufficient in natural resources, timber being a rare exception.[8] From quarries of limestone, sandstone and granite came pieces weighing from 2½ to 30 tons for the largest and oldest stone structures in the world. From them also were cut obelisks weighing several hundred tons each and at least one huge block of 1,000 tons.

For this quarry work, the Egyptians are supposed to have used only the simplest mechanical principles and appliances. The stone was marked off to the desired measurements; grooves were then cut with mallet and metal chisel or drilled. The pieces were finally split from bedrock by a process similar to what is known today as the plug-and-feather method. Bronze plugs were slid between thin metal feathers or wedges and driven in until the rock split, or wooden wedges were inserted which expanded when wet and split the rock. It was in this manner that the Egyptians are believed to have quarried limestone and sandstone. For their granite obelisks which were cut out horizontally, the Egyptians pounded the wedge with hard and tough dolorite, a coarse basalt, until they had cut a deep groove and finally broken off the piece they desired. This monumental building in stone of tombs, pyramids, temples and obelisks must have challenged the organizational and managerial genius of the men who built them.

The engineers and architects of ancient Egypt were exceptional men. They were among the first and the few, save royalty, to gain historic identity. They were appreciated in their own time and revered by succeeding generations. From the Old Kingdom, Imhotep, designer and builder of the first great pyramid for Netjerikhet, the Saqqara, was famous also as a physician ("magician") and maker of proverbs. His career exemplified the blend of superstition and reality characteristic of Egyptian life, preoccupied with hard daily labor and thoughts of the hereafter. It was said that his plans "descended to him from heaven, to the north of Memphis." His counsel was "as if he had inquired at the oracle of God." Two thousand years later, Imhotep himself was included in the pantheon of Egypt's gods.

Sennut, of the Middle Kingdom was Queen Hatshepsut's chief architect and engineer. He is best remembered for erecting her only surviving obelisk, which upon placement seated off-center and was slightly askew.[9]

Ineni, "chief of all works in Karnak," built principally in the New Kingdom. He came earlier in the distinguished line of engineers, but deserves the honor of the last place here for his own estimate of himself. "I became great beyond words," he said. "...I did no wrong whatever." Besides that, he was "foreman of the foremen." And, still he "never blasphemed sacred things."

SURVEYORS AND MATHEMATICS

Centuries before the Greek Pythagoras demonstrated the generalized relationship among the sides of a right-angled triangle, Egypt's "rope stretchers", or surveyors, are said to have been applying the knowledge that the angle between two sides of a triangle is a right angle if the sum of their squares is equal to the square of the hypotenuse. A papyrus now in the British Museum shows that they understood also how to calculate the contents of solids and to determine the slope or amount of cutback necessary in terms of the height of a pyramid and the length of its side. Three other facts regarding their mathematical notation are of interest. They used a system based upon 10, like ours, their fractions always had the digit I for the numerator, except for the fraction 2/3, and in finding the area of a circle, they used 3.16 for the value of pi (the ratio between the circumference and the diameter of a circle). The generally accepted approximation for this ratio today, for most calculations, is 3.1416.

Egyptian signs for their numerals were as follows:

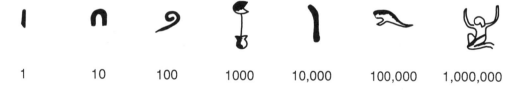

| 1 | 10 | 100 | 1000 | 10,000 | 100,000 | 1,000,000 |

For 1, a vertical line; for 10, an upside down u; for 100, a spiral; for 1,000, a lotus plant; for 10,000, an upraised finger; for 100,000, a tadpole; and for a million, a kneeling genie.

Fractions were represented by the combination of the sign for "month", representing 1, with the denominator written below it:

| 1/3 | 1/5 | 1/6 | 1/10 | 1/100 |

The eye of the hawk god, Horus, came to be used to represent measures of volume. The divisions were:

| 1/2 | 1/4 | 1/8 | 1/16 | 1/32 | 1/64 |

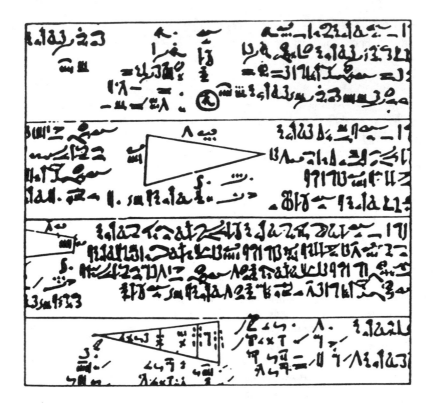

FIGURE 2-7.   Mathematical document. (Redrawn from Ifrah, 1985.)

Numerals based on the pictographic Egyptian hieroglyphs were not well-suited to rapid calculations and over time the users simplified them into a cursive system termed hieratic script. In its final form, the Egyptians used 36 signs for individual numbers. The change from the older system is illustrated in these examples.

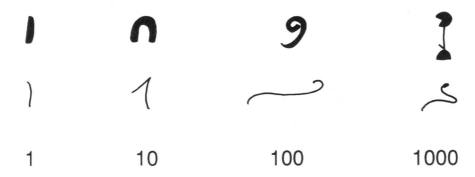

An important mathematical document dating from the 17th century B.C. is shown in Figure 2-7. It is a copy of an earlier treatise (ca. 1991–1786 B.C). Without a knowledge of the script, it is easy to grasp the subject matter as right triangles. With such tools, Egyptian builders can be more readily appreciated as skilled designers of their impressive constructions than as simple overseers of large labor gangs who happened to "get it right".

PYRAMID AND TEMPLES

Around 3200 B.C., predynastic Egyptian society witnessed continual regional power struggles until Narmer succeeded in unifying the valley which the ancient Egyptians called

FIGURE 2-8. Step Pyramid, Saqqara. (Library of Congress, 1971.)

Tawy, the "Two Lands" of upper and lower Egypt. Burials of Narmer and his predecessors, such as the legendary "Scorpion-king", were in large mud-brick tombs recessed into the ground called *mastabas*. They derive this name from their similarity to the shape of the low bench in front of the farmer's house called a mastaba. The outside walls of these underground tombs sloped inward at an angle of about 72°, or a tangent of 3 (71.6°). This slope of three to one (3:1) was carried over into later pyramid construction, although buttress or mantle walls required a less sharp angle of slope. At the Meidum pyramid, the slope utilized for the mantle resulted in a catastrophic collapse as we shall see later in this section.

The man generally credited with the invention of the pyramid was Imhotep, master builder for the Pharaoh Netjerikhet, also called variously Djoser, Joser or Zoser. In the tomb designed for Zoser (we shall use this name for no particular reason other than brevity), Imhotep utilized stone rather than mud-brick in the grand mastaba originally envisaged by these men. The shape and size were unusual as well, 70 meters square and 8 meters high. Typically, mastabas were rectangular and not as large.

This tomb underwent numerous changes. First, it was doubled in size, then again, and then a four-level step pyramid began to rise. In profile, the structure resembles several mastabas stacked one on top of the other (Figure 2-8). Still unsatisfied with the original four levels, Imhotep expanded the size of the design to a height of almost 70 meters with six levels (Figure 2-8). The step pyramid of Saqqara is located west of present-day Cairo. In ancient times, this was the site of Memphis, capital of the newly united Egypt.

It is with the next pyramid that Egyptian engineers learned a hard lesson in the construction of new designs. At Meidum, 56 kilometers south of Memphis, Qahedject (2608–2584 B.C.), last pharaoh of the Third Dynasty, began construction of another pyramid. Originally conceived as a step design, the structure was subsequently modified to a true pyramid by its builders.[10] The weight of the additional casing created compressive and shear

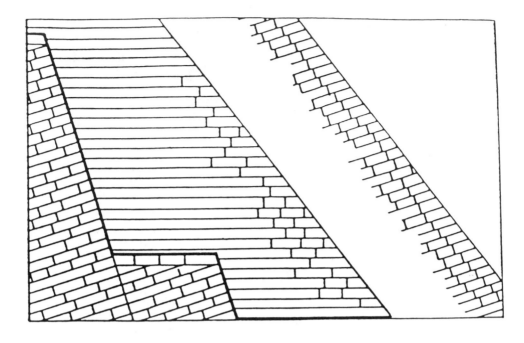

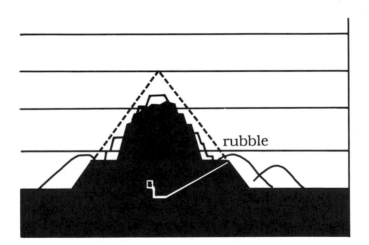

FIGURE 2-9.   *Upper:* Position of building blocks in the Meidum Pyramid. In this diagrammatical sketch the width of the unsupported mantle surrounding the step structure of the Meidum Pyramid is apparent. *Lower:* Meidum Pyramid. (Redrawn from Mendelssohn, 1971.)

forces significantly greater than that of the average pressure for a large pyramid. The average pressure at the base of a large pyramid is on the order of 50 k/cm^2. This is not great when compared to the strength of limestone and cannot cause failure if evenly distributed,[11] as with the inner, stepped core (1) of the Meidum pyramid (Figure 2-9, lower). The problems at Meidum began when the design was altered to include a second layer (2) or casing to which was joined a third layer or mantle (3) in order to achieve a slope of 52°.[12] The second layer's stones were laid on the original steps to smooth the structure's shape. Loads increased, but they were distributed by the underlying core. The 7 meter thick mantle's (Figure 2-9, upper) load raised pressures to 1,000 k/cm^2 at the joins of the blocks where compressive forces of this magnitude caused crumbling. This, together with the smoothed face of the second layer, provided a perfect situation where compression acts through a sheer unresisted by friction between the layers forming slip planes. Once failure started at Meidum, it must have been spectacular. We are left with no record of this event, only the uncompleted structure and the vast rubble pile of its mantle. By the next dynasty, the lesson had been well learned as the golden age of Egyptian pyramid building began.

During the Fourth Dynasty, in step with a new pharaoh, Khufu (Cheops), came the planning of a great pyramid that would be his tomb. Royal storehouses began issuing tools and clothing to the incoming laborers of the surrounding villages. These construction laborers were initially thought to have been slave workers, but are now believed, by Egyptologists, to be the peasants directly for the ruler. It is believed that these workers were drafted on a rotating basis for various months throughout the year. They were fed, housed and clothed as they served their god-king in a sacred cult of tomb building.

Between roughly 2700 and 2550 B.C., approximately 11 million cubic meters of stone were quarried and transported in forming pyramids, temples and causeways. All of the work was carried out without the use of beasts of burden or the wheel—not to include the use of rolling logs. Despite rigorous examination, the actual methods of construction used to achieve such a task remain somewhat mysterious. To better understand ancient construction capabilities, one must look at the builders' tools, equipment and methods or procedures.

Khufu's pyramid was constructed with more than two million stones weighing approximately two and one-half tons each. With the ancient construction methods available, the pyramid was still completed within the Pharaoh's 23-year reign, at about 2600 B.C. The engineers involved with its construction had to come up with a number of innovative solutions. The pyramid has an average length of the sides of 236 meters at the base with three or four minutes of error. Rods were used, graduated in cubits, palms and digits (an Egyptian cubit was a bit over 0.5 meter, a palm was 10 centimeters and a digit was about 2 centimeters).

The structure was carried upward at a uniform slope of 51°51', with its 206 courses reaching a height of 146 meters. It is a gigantic pile of $2\frac{1}{4}$ million rough-looking, but carefully squared and placed, limestone blocks.

The interior of the Great Pyramid is, like that of the others, mainly solid masonry. There is, however, an inclined passage, large enough for a man to crawl through, that leads inward and downward from the north face at nearly $26\frac{1}{2}°$ with the horizontal. Branching from it is another, leading upward into a tall but narrow "great gallery" through which one reaches the "king's chamber", nearly, but oddly enough not quite, under the apex of the pyramid. This chamber, 5 by 11 meters, which Khufu trusted would hold his sarcophagus inviolate forever, is lined with granite and roofed over with corbeled granite slabs. The roof is a succession of five large relieving chambers obviously planned to take part of the enormous superincumbent weight of more than 100 meters of blocks.

Surveyors first marked the base to form a perfect square. The next step involved the construction of the multiple-step, layered terraces on which the pyramid was formed. The terraces had to be perfectly level to prevent the structure from being skewed. To achieve this, the builders erected a large system of water-filled trenches about its base. They used the water to level the 5 hectare site to within a half-inch from one corner to the opposite diagonal corner.

It was first believed that 100,000 men worked on the pyramid but there were probably only a fraction of this number working at any one time. This figure is based on an analysis of the area of the barracks used by the permanent staff of about 4,000. Further study of tomb paintings indicate 172 men could move a 60-ton statue, so eight men could move an ordinary 2.5-ton block. Additionally, oxen could be used as well. Gangs that hauled the heavy stones up ramps and set them in place were composed of approximately 18 to 20 men. Although the work was extremely tiresome, it is reported that the gangs were pleased to work for the pharaoh. Working conditions were favorable, with men never getting exhausted or thirsty. The blocks, after quarrying, were painted or marked with a destination or labeled, ''this side up.'' The block inscriptions also made known the existence of workmen team identities, which displayed the team's pride in their accomplishment. Such inscriptions read, ''The Craftsmen Gang — How powerful is the White Crown of Khnum-Khufu,'' ''The Victorious Gang,'' or ''The Enduring Gang.''

The ancient Egyptians never lifted stone blocks by means of pulleys or suspension by ropes. The stones were mobilized through jacking operations utilizing wedges, levers or rockers. Jacking involved exerting a considerable amount of force with very little movement, which was taken advantage of due to the limited boundaries within which to work. Levers exerted a great amount of pressure when the fulcrum was placed close to the stone. These levers had an additional counter-balance with a weight on the opposite end of the lever. The lever was used horizontally for lifting and vertically for lateral movement. The rocker was used for lifting medium-size stone blocks. It was a two-lever type of assembly which rocked the stone back and forth, while wedges were placed under the stone in the lifted position. This enabled the teams to lift the blocks quickly and safely.

The technique for lifting medium-size stones was inadequate for large blocks. For these, a jacking device appears to have been the more practical means of transporting the stones. One such device was the balance beam, which was simply a free horizontal beam placed on a fulcrum which was set beneath the beam near one end. The beam pivoted about, with the short end acting as the main leverage. It allowed for close control throughout the lift. With long and multiple levers, no stone block was too big to lift onto a sled. Because the balances were usually of timber, it is said that they were used to the limit of their endurance and then used for firewood.

It is now generally believed that the stones were transported mainly with the use of ramps. A 2-ton block could not be dragged through loose sand. The ramp surface was hardened either with stone blocks or chips from the quarries, or with a proportional mixture of wet sand, silt and clay. According to the straight-ramp theory, which proposes that the Egyptians built a sloping roadway up to the level under construction, layer upon layer was superimposed upon the original slope as the pyramid rose. Each layer, therefore, required surfacing. Huge granite lintels, like those carrying the roof of the king's chamber in the pyramid of Khufu (Cheops), were raised, perhaps, by ''seesawing''. Each stone, weighing roughly 55 tons, could have been made to rest on two supports close to its center. It then could have been rocked back and forth and raised an inch or two at a time by inserting small blocks on each support alternately. A well-drilled gang of men walking from one end to the other might get the block as high as 7 meters in the course of a day's work.

The major ramp, bearing most of the construction traffic, led from the river bank to the job site. Some believe that most of the stones were transported to their designated terrace by ramps. This thought, however, has been dismissed for the upper levels of the pyramid where space was limited for the slope needed. Due to the vague ideas of how the upper levels were constructed, its construction details are avoided somewhat and not generally discussed. An additional reason for the use of the access ramps was to facilitate the transporting and maneuvering of the oversized blocks of the king's chamber along with the 50-ton granite plug, the great tiers of ceiling beams, and the pairs of tilted relieving stones above them. Above the elevation of the ramps' termination, stones could be placed by being rocked up in step sequence with the exception of the pyramidal capstone. The capstone is believed to have been stepped up at the completion of each terrace until final completion of the pyramid.

The finishing touches of the pyramid included placement of the cornerstones on the diagonals and the smoothing or polishing of the planes. These corner diagonal stones went through a type of tensioning or jacking system in which a device known as the Spanish windlass was used. The cornerstones had two pegs driven into the outer shell, allowing a rope to be looped around. The other end of the rope was secured to the center large stones in the middle of the pyramid. Following the placement of the cornerstone on the terrace and the attachment of the rope at both ends, a stick was twisted between the laps of the rope, thereby producing a tourniquet effect which caused the rope to tighten. This procedure nudged the stone securely against the adjacent planes of the particular corner. It simply jacked the stone into the rest of the structure. It also served as a method to align the sloping edge of the corner block with the sight lines established for the diagonals of the pyramid.

A final note concerns the mortar used. It consisted of sand and gypsum with other admixtures of impurities. It had no adhesive strength; however, it was not used for that purpose. It was used for two other important reasons. One was to insure that each stone rested evenly and completely on the blocks below. It aided in obtaining an even distribution of weight on each stone by filling any voids and cracks. This could be critical, as we have seen. Its second use was to facilitate the laying of the stones.

Not all the methods discussed above for the construction of a pyramid may be correct. It really is not known whether the Egyptians used the balance beam or the rocker, or some other method.

Areas which are not understood will always lead to some unusual theories. One such theory is that of a French chemist named Joseph Davidovits.[13] He contends that the stones of pyramids and other ancient structures were not hauled in by teams of peasant workers, but rather were formed on the project site. He had discovered a way to form rocks, which the ancient engineers might have known. The chemistry requires aluminum and silicon of common clays, and a proper amount of alkali, which could be obtained from the water of the Nile. This mixture forms a molecular glue. When some type of crushed stone, which can be easily transported with buckets, is mixed in and then put in a form under low heat (the sun was adequate) a stone block could be formed. Interesting as the idea is, the archaeological evidence militates against this theory. The finding of copper, saws, wooden mallets and chisels in the huge ancient quarries indicates the stones were cut rather than molded using a lost geopolymer science.

Beyond the pyramids, the Egyptians continued their building in stone. The Middle Kingdom temple of Amon-Ra at Karnak was once, if no longer, the largest columnar structure in the world. Its dimensions of 103 by 372 meters are sufficient to contain the combined ground areas of St. Peter's at Rome, the Cathedral of Milan, and the Notre Dame at Paris. The great hall of the temple, known because of its rows of columns as the Hypostyle Hall, was 100 by 52 meters (329 by 170 feet). The columns stood 21 meters (69 feet) high in the center aisles and 13 meters (42 feet) along the sides. They were more than 3 meters (10 feet) in diameter, supporting short architraves—or crossbeams—that weighed 60 to 70 tons each. These carried the flat roof at two levels, making room for clerestory windows, apparently the first in history.

To place a column of several hundred tons upright is an engineering feat that requires nicety of calculation and special equipment even in modern times. The length, weight, slenderness and relative weakness under bending stress of such huge blocks of stone make the task uncertain at best. It is obvious that Egyptian engineers solved the problem successfully though, as at Stonehenge, it is not clear just how.

Egyptian masons applied their simple instruments with ingenuity and care. To insure precise and firm bedding, they used mortar made of gypsum with very little sand. Such precision, however, was customary only in the outer veneer where the surface was tooled and dressed after the stone had been laid. Inside work was apt to be coarse, with more sand in the mortar. Walls of two faces were often built without connecting stones to hold them together, and sometimes the space between was filled with loose rubble. This observation is not an indictment of the Egyptian builders, as it was used by the later Romans and is still found in use today.

EGYPTIAN MILITARY ENGINEERING
Buhen

Egyptian building in the Nile Valley below the First Cataract was typically non-military. The record in stone reflects a concern with monumental works such as the pyramids of the Old Kingdom, temples—both mortuary and public—and other public buildings. Fortifications were present in the form of walls such as at Memphis, but were not elaborate designs and consisted of one rampart. Imperial expansion of the Middle and New Kingdoms led to the development of Egyptian fortress engineering and as with most Egyptian works, they built well.

Beyond the first Cataract in Nubia, the Egyptians built eight fortresses and beyond the second Cataract, they established six. The best preserved, until the construction of the Aswan High Dam, was Buhen, southernmost of the first eight fortresses. Buhen was the administrative center for the Cataract area. It was built with one side along the Nile and was provided with docks. The walls of the remaining three sides were built with two ramparts (see section below) with a moat. The counterscarp opposite the first rampart had a brick-faced glacis. The first rampart was built with circular towers at intervals. These towers had loopholes arranged so that archers could traverse the entire moat and glacis. Each bowman had a 60° field of fire (Figure 2-10). The second rampart was 10 meters high and provided with square projecting towers. The wall thickness of the interior curtain was 7 meters at the plinth and 5 meters at the top. The corners were all towers. The masonry was of mud brick reinforced with timber on a base or rock and rubble fill. Buhen was built in about 2000 B.C.

It cannot be said that the Egyptians made any great innovation or departure in the technique of building with stone. They reproduced what had been made earlier in wood, mud or brick. Their pyramids were but magnified mastabas, block upon block above a sunken chamber tomb. Great as Karnak and other temples were, they never got beyond the post-and-lintel form. True arches, made of wedge-shaped mud bricks, were known in Egypt, but were never brought aboveground and adapted to construction in stone.

A final Egyptian feat of engineering was a canal to connect the Nile with the Gulf of Suez, thus linking the Mediterranean with the Red Sea. It may have been begun around 1870 B.C., but according to some authorities the project was abandoned. Others think that the canal was in use during the reign of Queen Hatshepsut, who died in 1468 B.C., and that it later silted up. There is a picture of a trading expedition in her time down the Red Sea to Punt, now within the territory of modern Ethiopia.

Later Mesopotamia — Assyria and Babylon

The Assyrians came from a city-state on the upper Tigris River known as Ashur (Figure 2-11). The first important Assyrian ruler was Ashuruballit I (1365–1330 B.C.), who incorporated prime corn-growing areas of northern Assyria into a new empire (Figure 2-12). His successors, Shalmaneser, Ashur-nasirpal, Sargon II and Tiglath-Pileser, expanded the empire by military conquests.

Ashur-nasirpal (883–859 B.C.) built his capital in Nimrud with 16,000 inhabitants. Sargon II (722–705 B.C.) moved the capital to Khorsabad (Figure 2-12). Sennacherib (704–681 B.C.) succeeded Sargon II, with the capital located in Dar Sharrukin (Khorsabad). This city was one of the first to use city planning with crossing right-angle streets and large squares. To the new king, it was a city in a desert and he wished to have greenery. To have this, he set about monumental hydraulic projects. He sited a new capital at Nineveh, two

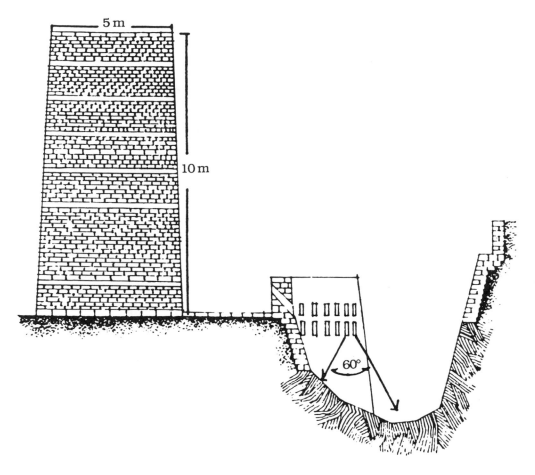

FIGURE 2-10. Buhen—section. (Redrawn from Sandström, 1970.)

miles up the Khosr above its confluence with the Tigris. The city was built in two years complete with palace and curtain wall. A park surrounded the palace and a broad green belt reserved for fruit orchards surrounded the city. To irrigate the parks and orchards, Sennacherib straightened and canalized the Khosr river 16 kilometers upstream of Nineveh. Dikes raised the river level to permit irrigation around Nineveh. To expand this regime of irrigation and "greening" of the desert, by 700 B.C. Sennacherib had built dams north of Nineveh and joined three rivers 20 kilometers northeast of Nineveh to his canal system (Figure 2-13).

In 690 B.C., Sennacherib began a 237-kilometer canal from the northern Tas Mountains which included the great 280-meter aqueduct at Jerwan. This aqueduct was 20 meters wide. Corbeled arches were used to cross the river in the valley. The completed structure was 7 meters high and contained about two million stone blocks dressed to 50 × 50 × 50 centimeters. The 80-kilometer Bavian canal, of which the aqueduct was a part, was completed in one year and three months.

The canal itself was built with a 40-centimeter (16-inch) bed of concrete poured on a 1-inch bed of mastic with a stone pavement finely jointed. The bottom had a gradient of 1:800 and the concrete used was one part lime, two parts sand and four parts limestone aggregate.

The "mastic" was bitumen mixed with fillers, generally obtained at Hit on the Euphrates 320 kilometers north of Babylon. This was a seep source and free of adulterants. It was by

FIGURE 2-11.   The city of Ashur on the Tigris, in the Middle Assyrian period, as seen from the northwest. A drawing by Walter Andrae, toward the end of his ten-year excavation. (Lloyd, 1978.)

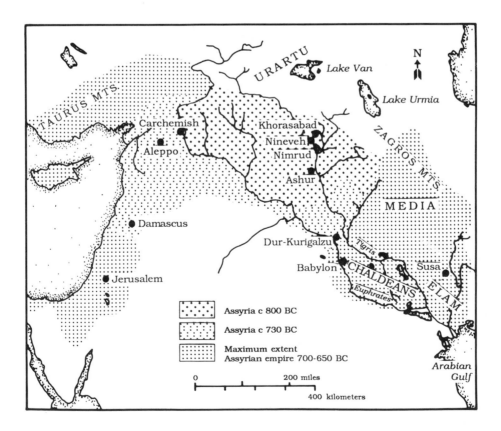

FIGURE 2-12.   Map of the Assyrian empire. (After Postgate, 1977.)

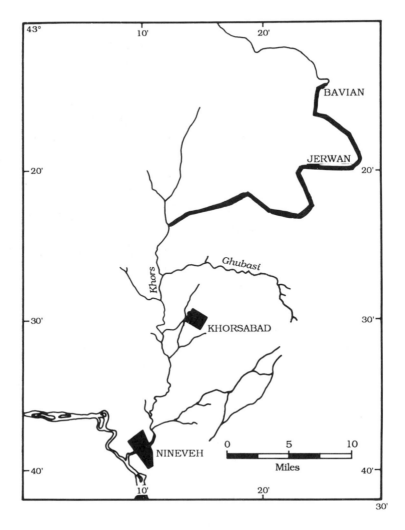

FIGURE 2-13. Map showing the line of Sennacherib's Canal, from the barrage and sluices at Bavian, passing over the Jerwan aqueduct and finally joining the course of the River Khosr to Nineveh. (After Lloyd, 1984.)

weight at 3¹/₂ shekels of silver/1000kg. The mastic waterproofed homes, buildings, canals and even barges. This mastic was utilized extensively by Nebuchadnezzar (604–561 B.C.) in building a 26-kilometer (16-mile) high dam joining the Tigris and Euphrates at Babylon. It was an earthen-filled dam lined with kiln-fired brick set in the mastic.

Sennacherib said: ''I caused a canal to be dug to the meadows of Nineveh. Over deep-cut ravines I spanned (lit., caused to step) a bridge of white stone blocks. Those waters I caused to pass over upon it.'' Even more astonishing, he gave credit to his engineers and men. A large relief made of alabaster and placed on a wall of Sennacherib's palace in Nineveh shows how heavy statues were transported in mountainous country.

Assyria had a special corps of men called *ummani* to level the ground for baggage carts and to build temporary bridges. The roads which they constructed must have served for commerce and travel in times of peace as well as war. Northeast of Nineveh, a portion of Sargon II's great palace, built in the 8th century B.C., still stands. Originally, it was composed of three groups of buildings extending over 10 hectares and containing 200 rooms. Its walls were of moist or partially baked bricks that stuck together as they dried and hardened. They were faced either with stucco or enameled brick. But in some lower parts of the walls, there are great limestone monoliths, one weighing more than 20 tons, the architectural and engineering purposes of which are not apparent. The courts of the palace were elaborately paved with stone set in asphalt. There were also stormwater drains emptying into brick-lined sewers, which led to main conduits covered with flat slabs of stone or brick vaults.

The last of the great Assyrian kings was Assurbanipal, who died in 630 B.C. Political chaos followed his death and by 612 B.C. the Persians and Babylonians had sacked Nineveh, the capital built by Sennacherib during his reign (705–681 B.C.). The perimeter walls totaled 80 kilometers (50 miles). This curtain was 40 meters (120 feet) high, 10 meters (30 feet) wide, and was equipped with 1,500 towers. The wall was brick-faced and rubble-filled. Nineveh stood as a fortress city for less than a century, falling in 612 B.C.

The city was occupied by a combined Scythian and Babylonian army of 400,000 men for over two years. Unable to breach the walls by sapping or rams, the besiegers may never have taken Nineveh but for an inundation of the city wall foundations by flood. Slumpage due to erosion and infiltration by flood waters created a breach 20 stades wide (approximately 3,500 meters). The attackers quickly exploited the breach and captured Nineveh. Assyria fell with it.

Babylon

Hard-surfaced streets were early achievements in engineering. Men walked on cobble-stones in Assyria 4,000 years ago and there were brick and limestone pavements later in Babylon. Babylon, the city, was the seat of two Mesopotamian empires ruled over by its kings.[14] Both were relatively short-lived as empires go, being the creations of two resourceful and powerful men, Hammurabi (1750–1708 B.C.) and Nebuchadnezzar II (605–562 B.C.). With the transience in nation-states and ''empires'' in the modern world, it seems somewhat misplaced to speak of any ethnic or cultural continuity between periods separated by a millennia. Still, this appears to have been the case in Babylon and, indeed, in both Meso-potamia and Egypt. City-states and kings could gather varying amounts of power and territory unto themselves, but the underlying civilization remained relatively the same no matter what face it wore—Sumerian, Amorite, Elamite or Chaldean. Only the Assyrians seemed to diverge from this model, yet even they gave deference to Babylon to the extent that its outright defiance of Assyrian rule led to reprisal and destruction only once, in 691 B.C., during the reign of Sennacherib. Even so, Esarhaddon, who followed his father, Sennacherib, rebuilt Babylon. After the fall of Nineveh in 612 B.C., Babylon reached its greatest height. It is this Babylon that is written of by Greek authors such as Herodotus.

This new Babylon was laid out as a rectangle bisected into two parts by the Euphrates (Figure 2–14). Around the perimeter were its famous walls, estimated at 365, 385 or 480 stades (roughly 80 kilometers) built to a height of 200 cubits (95 meters) and a width of 30 cubits (15 meters). The battlement or outer parapet was crenelated in a square zigzag pattern familiar to us in much later medieval fortresses.

Its famous Processional Street, leading to the bridge across the Euphrates, had a foundation of asphalt-covered brick and limestone flags with beveled joints set in bitumen and mortar. The roadway was surfaced with large stones, one meter square. The sidewalks were of smaller slabs of red breccia, a sort of conglomerate. Seven pier foundations of the bridge have been found in the river bed. This ancient highway drawbridge was constructed in the time of the Second Babylonian Empire (612–539 B.C.). Its piers were made of small burnt bricks, with each pier measuring 10 by 20 meters. They were shaped with the pointed end upstream to cut the force of the current, and the end downstream rounded to reduce scour currents which would undermine them.

The large triangular salient formed by the outer wall on the east side of the river enclosed the residential area of the city. Over 20 meters high, the most famous gate of the city, the Ishtar Gate, was a vaulted passage finished in enameled brick—blue on the towers, green

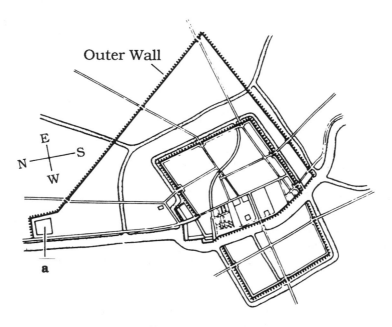

FIGURE 2-14. Babylon at the time of Nebuchanezzar II (605–562 B.C.) with fortifications extended to enclose the Summer Palace (a) in the north. (After R. Leacroft, in Lloyd, 1978.)

and pink on the connecting walls. Reliefs of bulls and dragons decorated the towers and walls leading to the Processional Street.

The ziggurat most remembered by modern men was the so-called "Tower of Babel". To the ancient Babylonians, it was the Etemenanki, the "Cornerstone of the Universe". Rebuilt several times, it reached its final form under Nebuchadnezzar II, towering to nearly 100 meters and covered in enameled brick and reliefs. It has a base area of nearly 7,000 square meters.

For all its magnificence and fortress engineering, Babylon only briefly survived Nebuchadnezzar. Nebonidas, the last monarch of Babylon, was no scion of his nobler ancestors. In 539 B.C., Cyrus II of Persia occupied Babylon, ending the Mesopotamian three millennia self-rule by one city or another.

ANCIENT SHIPS — BEAMS AND TRUSSES

The earliest boat design was that of a modified tree or log which probably evolved into a dugout. Reed and skin boats have been suggested as the early Mesopotamian vessels. Egypt's early boats are illustrated on sandstone rocks and they appear as reed boats. These quickly evolved into vessels built of planks or strakes. These early planked boats were built without ribs or frames. Sail and rowing was utilized for propulsion in the tideless Mediterranean Sea. A pot dated to 4000–3000 B.C. indicates a square sail with yard and boom. The sails of these early craft were fragile and several lifts or support rings on the boom were necessary to relieve stress. Steering oars were first developed to aid sail craft. The Bronze Age has been termed the "First Great Age of Seafaring."

Egyptian vessels utilized bipod or "sheerleg" masts. This was due to the weak hulls of early boats, which could not bear the compression of a single mast stepped on the centerline where the unbraced or little-framed hull was the weakest. The double mast placed the stress on each half of the hull. The ships of the 25th century B.C. had multiple steering oars and propulsion oars for moving into the wind, and the multi-stayed, hinged bipod mast with a fixed boom. The yard was held by stays or vertical tensioning lines. A hogging truss was utilized and by the 16th century B.C. a fixed boom was replaced to allow closer sailing to the wind. The yard was controlled by "braces", the boom by "sheets", (fore-and-aft tensioning lines). Due to the sail arrangement, sailing with the wind more than 40° forward of dead astern was not feasible.

Early hull design by the Egyptians was predicated on the dugout log and was the basis for the smooth-planked carvel hull. Boats of this type are suspected to have existed in the Archaic Period between 3100–2700 B.C. Cedar wood, imported from Lebanon, was worked with copper, later bronze, adzes, chisels, awls and saws. Ropes were made of flax of halfa grass with sails of linen. The best example of Egyptian shipbuilding is the Boat of Khufu (Cheops) excavated from a boat grave near the great Pyramid of Khufu (Cheops). In cross-section, the boat has no keel but has three broad, heavy bottom planks fastened edge-to-edge. The decking is supported by flat transverse beams which articulate with the sheer plank via holes drilled into the latter. On the centerline, directly under the deck beams, was the Egyptian substitute for a keel—a long, continuous wood girder set on edge and notched to all the deck beams to cross over flush. This girder ends just short of the bow and stern. It is supported at regular intervals by column stanchions resting on transverse foundation members or cross-framing. These cross-frames or floors are more akin to modern vessel structural members where these are supporting bases rather than frames in the sense of providing lateral resistance to external stresses created by water pressure on the hull. The longitudinal girder thus functions more as a centerline truss than as the rigid beam utilized in later ship design.

This emphasis on longitudinal strength in hull design anticipates a radical conceptual shift in modern steel hull design in the VLCC (very large cargo carriers) and ULCC (ultra-large cargo carriers). This rather complex truss system was absolutely necessary to prevent the phenomenon of "hogging" or opening of hull seams due to the sinusoidal motion of a vessel in a wave train with a wavelength equal to or greater than the hull itself. The hogging phenomenon has its opposite in sagging, where hull seams are compressed at the upper gunwale and deck structural members while lower members are placed in tension. This truss design was needed in the hulls used by the Egyptians with length-beam ratios of up to 7.2:1 in the Khufu (Cheops) Boat 42 meters:6 meters. The weakness in Egyptian truss construction was the fasteners available to the builders. These fasteners were rope fiber or heavy twine laced through holes in adjacent wood structural members. With stress and moisture, such fasteners became fatigued and loosened, resulting in short use-lives for the vessels. Efforts to control the slack in their vessels began a few hundred years after the Khufu (Cheops) vessels.

A hawser-like bridge appears in 2500 B.C. attached at each end with a girdle or loop around the bow and stern. The vertical support stanchions provided a deeper structural web and rested on the floor frames. This hawser was equipped with a tourniquet rod thrust through the strands for periodic tensioning. This anti-hogging truss was used for over 10 centuries while the internal heavy deck girder under decks also continued in use. One can suppose the bridge truss design was more appropriate to seagoing craft where wave activity would be greatest and the girder truss used in river navigation on the Nile. It was not until 1500 B.C. that we see evidence of a keel in Egyptian vessels and it was probably not an Egyptian development.

NOTES

1. **Wittfogel, K. A.** *Oriental Despotism: A Comparative Study of Total Power.* Yale University Press (1957).
2. **Steward, J. H., Ed.** *Irrigation Civilization, A Comparative Study.* Pan American Union (1955).
3. **Redman, C. L.** *The Rise of Civilization.* W. H. Freeman, San Francisco (1978).
4. **Mitchell, W. P.** The hydraulic hypothesis: a reappraisal. *Curr. Anthropol.* **14**(5): 532–534 (1973).
5. **Friberg, J.** Numbers and measures in the earliest written records. *Sci. Am.* **17**: 110–118 (1965).
6. **Ifrah, G.** *From One to Zero.* Viking Press (1985).
7. **McEvedy, C.** *The Penguin Atlas of Ancient History.* Penguin (1967).
8. **Etchwerry, B. A. and Harding, S. T.** *Irrigation Practice and Engineering.* (2 volumes). McGraw-Hill, New York (1933).
9. **Malek, J.** *In the Shadow of the Pyramids.* Golden (1986).
10. **DeCamp, L. S.,** *The Ancient Engineers.* Ballantine (1963), p. 40.
11. **Malek, op. cit.,** pp. 44, 45, 69.
12. **Mendelssohn, K.** A scientist looks at the pyramids. *Am. Sci.* **59**: 210–220 (1971).
13. **Davidovits, J.** Ancient and modern concretes: what is the real difference? *Concrete Int.* **9**(1): 23–29 (1987).
14. **Lloyd, S.** *The Archaeology of Mesopotamia.* Thames and Hudson (1978).

3 CLASSICAL ANTIQUITY I: THE GREEKS

Of all ancient civilizations, none has been written of more than classical Greece at her height with such immortal names as Pericles, Socrates, Plato, Aristophanes, and Democritus. While this period burns most brightly in the continuum that was Greek civilization and culture, its importance takes on greater significance when considered within that complete sweep of history. Just as the Old Kingdom pyramid builders represent a significant point in the history of Egyptian engineering, we must consider Classical Age Greece and her world within the overall picture of Greek cultural history. To do so gives us a better appreciation of the origins of modern scientific principles and thought that so influence modern engineering practice.

In a history of engineering, science or technology, Greece presents an enigma. For all her brilliance in science and philosophy, it is difficult to find numerous tangible examples of applications and certainly not on a scale approaching that of earlier or later cultures such as the Egyptians or Romans. There were the well-known "wonders": the Colossus of Rhodes, the Pharos or lighthouse of Alexandria, and, of course, the Parthenon. Only a mature engineering sense could have produced these famous works. Of these examples only the Parthenon remains. It is more of a tribute to architectural skill than the others, which vanished long before the modern era.

The practical application of learning was not the imperative in the ancient Greek mind that it is for us today. Greek concepts of science and technology were strongly influenced by philosophers such as Pythagoras and Plato. These were arts to approach with a "liberty of mind" that is liberated from a concrete construction which embodied their principles. It is one thing to conceive of the principle of the lever and quite another to build and demonstrate one. This emphasis on abstraction seems to create a paradox in evaluating Greek accomplishments in engineering. Clearly, they understood and, indeed, originated or improved many of the principles used by later engineers. Technology, to the Greek mind, was mechanization; those who applied its principles were "mechanics" or "mechanikoi". The later Roman author, Seneca, bluntly described them as "workaday mercenaries". This rather extreme connotation did not, however, produce the society of philosophers and slaves envisaged by Plato's *Republic*. Greeks were practical as well as philosophical and became even more so after the Classical Period (600–400 B.C.). We shall consider now the antecedents to that period in the Greece of Homer—Mycenae.

HEROIC GREECE — MYCENAE

The Mycenaean period of Greece is that which comes down to us in the epics of Homer. He, while a poet, also plays the role of ancient historian in that his writing occurred at least five centuries after this period. The Greeks of Mycenae, who were in possession of the peninsula by 1600 B.C.,[1] are more properly termed "Achaeans". At sites like Mycenae and Tiryns (Figure 3-1), the predecessors of Homer's Agamemnon built great palaces and walled citadels of heavy masonry. Unlike other Mediterranean cyclopean towers of the neolithic Sardinians called "Nuraghes", these structures were built for both defense and habitation. They made extensive use of the corbelled arch in tomb galleries and wall casemates. By rotating the corbelling through 360°, they achieved domed chambers in their royal tombs or "tholai" (Figure 3-2).

This late Bronze Age society left few records of their builders. Their written language was an early Greek script termed "Linear B", which was modified from "Linear A", a

FIGURE 3-1.   The Aegean area. (Redrawn from Kirby et al., 1956.)

script written by Kings of Minoan Cyprus during the Mycenaean Period.[2] At the Cypriotic royal sites of Knossos and other palace ruins, we can appreciate the skill with which these builders used stone and wood in their constructions—arcaded facades, terraced staircases, windows, and interiors whose pillars opened space rather than enclosed it like the Egyptian, Karnak. The Minoan variant of Mycenaean civilization shows a less monumental treatment in construction. While impressive, these structures call upon people to enter and enjoy them rather than being simply dim sanctuaries of god-kings. In the palaces of Crete, Minoan engineers moved water through terra cotta pipes that were tapered to increase velocity to keep downstream portions clear of sediments.[3] Runoff from these pipes, after descending in parabolic-shaped gutters carefully designed to prevent too great a flow velocity and, hence, splashing or sloshing-over losses, were collected in cisterns. Settling basins collected sediment, while sunlight helped keep the water pure in its passage to the storage cisterns.

Minoans borrowed from Egyptians with whom they traded over eastern Mediterranean sea routes. Their palaces, however, owe more to the style of Karnak than that of Mycenaean Greece. This is exemplified by the total lack of the corbelled arch in Minoan building. Ruins of the causeway from Knossos to its port are an example, where great ashlar piers carried the roadway on wooden beams rather than on stone arches. The road itself represents an appreciation by those early engineers of a solid road base, along with drainage and paving. Their road foreshadowed Rome's stone paved *vias* of 1,000 years later.

The road was founded on rough stones embedded in subgrade. Smooth stone slabs then set in clay cement and meter-wide shoulders of a rammed gravel-pottery aggregate in clay made a total width of three and one-half meters. Unlike later Roman roads, it had only one drain and no camber to its surface to aid in runoff. No evidence of wheeled traffic has been found and these roadways have been termed "processional ways". Whatever their function, they represent one of the few examples of engineering we have from this period. We cannot call the Achaeans or their Minoan cousins true Greeks, but the germ of Greek style is there at this early time.

By 1100 B.C., Mycenae, Knossos and many other centers had fallen. Mycenaean Greece's last deeds were perhaps the reason we know of them at all. Homer brings them to us as the protagonists, along with the Trojans, in the fabled but real Trojan War. Troy, as at the Mycenaean centers, had the citadel architecture and stone walls that, for want of written histories, bespeak a world where conflict was common and prepared for by its people.[4]

In the following centuries, Ionian Greeks moved eastward to the Anatolian mainland coast while mainland centers such as Pylos lay ruined or abandoned.

CLASSICAL CIVILIZATION

After the "Dark Ages" (1000–700 B.C.) following the fall of Mycenae, what we know as classical Greek civilization arose in the form of individual poli or "city-states". The Greeks gave us the concept of our form of government (democracy), great works of art (sculpture, architecture), the basis for our philosophy, and science. In an engineering sense, Greek civilization gave us critical thought and concepts such as pi, zero and the square root. In comparison to the later Romans, these Greeks were by no means engineers. For historical perspective, Greece must be placed in the context of a 7th–6th century B.C. world.

Persia had conquered the Assyrians in 606 B.C. Under Xerxes, the Persians conquered the Illyrian colonies of western Anatolia and invaded Greece. The Battle of Marathon in 479 B.C. dealt Xerxes a major defeat, finalized by the naval defeat at Salamis in 467 B.C. Greece entered into its "Golden Age".

The Ionic philosophers of the Greek Illyrian coast were responsible for the beginning of scientific thought that is associated with the Classical Period. These include: Thales of Miletus (640–546 B.C.), who stressed empirical study; Pythagoras (582–500 B.C.), who developed abstract mathematics; and Democritus (500–420 B.C.), who developed atomic theory and influenced Hero of Alexandria.

GREEK CONCEPTS

Greek philosophers, notably Pythagoras and the Ionian, Thales, defined abstract rules for studying physical problems. Archimedes (287–212 B.C.), who followed Thales, developed technical mechanics following Euclid's method of logical deduction. Archimedes better fits our modern definition of scientist or engineer. He defined problems for himself and designed solutions, but also worked for clients (e.g., Grecian Syracuse's defenses against Roman siege). Archimedes represents a later Hellenistic view of Greek science, but the useful application of theory remains in opposition to Classical maxims.

The opposition of the "arts of liberty", or liberal arts, to engineered work is derived from Platonic and Pythagorean concepts of pure art and science which exalt philosophy above artistry. A general trend after the 6th century B.C., Thales stressed empirical study. Greeks such as Thales were concerned with "Static" rather than "Dynamic" (e.g., Idea is formless, timeless vs. change, motion). False opposition was never recognized by Classical Greeks. Technology, to the Greek mind, was machination, i.e., "an artful method" from the Greek, "mechanaomai—to create or contrive a deception."

As early as 2800 B.C., a hieroglyph translated as "Chief of Works" appeared in Egyptian records. By the time of the Greeks, their builders were called "architetkoii" and "me-

FIGURE 3-2a.   Mycenean tomb, Treasury of Atreus.

chanikoii'' (from ''mechanaomai''). These terms cannot be translated exactly as ''engineer''. The architekton was principally a technician in charge of construction and was often the designer as well. By the 5th century, these men worked under a contract system. The contract was often inscribed on stone and placed on-site. General construction details, supplied by the architekton and his craftsmen, would be listed.

EARLY GREEK ENGINEERING

In 600 B.C., Eupalinus of Megara constructed an aqueduct for Samos. The water was drawn from an inland lake on the island and passed through a tunnel 2.5 meters in diameter and a kilometer and one-half long under a hill 280 meters high. Within the tunnel, a meter-wide channel contained pipes for the water.

In the Samos Tunnel, two headings were used. The errors in grade and alignment were 1 meter and 7 meters, respectively. Hero of Alexandria (ca. 100 B.C.) describes an alignment survey (Figure 3-3) that could have been used in the Samos Tunnel. The technique would have used right-angle measurements following the most favorable ground, using the law of the square.

Thales had brought Egyptian geometry back to the Greek world. Thales took Egypt's practical geometry into the abstract by development of concepts where a line becomes a

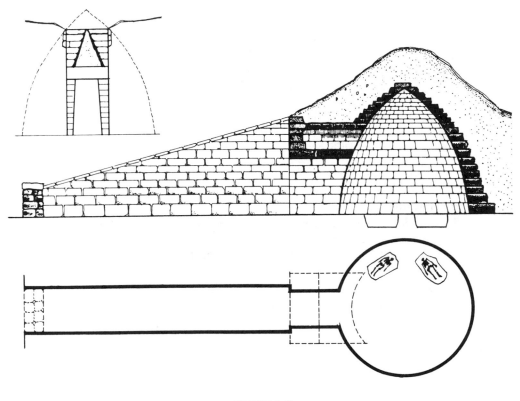

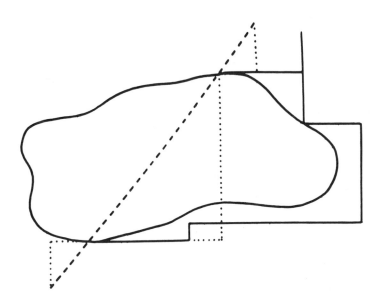

FIGURE 3-2b.

FIGURE 3-3. Tunnel alignment by right-angle geometry. (Courtesy of Professor O. A. W. Dilke.)

directed ray. He defined the equality of angles and extended the concepts developed with the circle to the line as shown in the following examples of Thales' Theorem of how to lay out a right angle (Figure 3-4). His theorem states that: "An angle on the circumference of a circle subtended by the diameter is always a right angle." In practice, the procedure is:

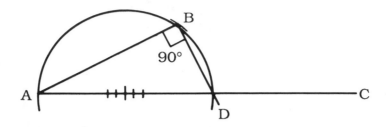

FIGURE 3-4.   Illustration of Thales' Theorem.

1.   Lay out side AB direction.
2.   Measure distance AB to establish B.
3.   Lay out auxiliary line AC.
4.   Find point on AC line that would serve as center of circle passing through A and B.
5.   Where this arc of circle intersects AC is D.
6.   Joining of B and D, the right angle ABD is formed.

At Samos, the Ionian engineers also built a half-kilometer long breakwater in 30-meter water. A diolcos, or slipway, was built 5 kilometers long at the Isthmus of Corinth for the movement of vessels or cargoes. This causeway for ships allowed the Greeks to avoid a journey of 720 kilometers around the Peloponnesus (Figure 3-1). The vessels literally were portaged from one shore to the other, using cradles on rollers. Unlike later seagoing vessels, these Greek ships were designed to be beached rather than anchored and were loaded and unloaded on ramps much like the modern motorboat.

Generally, the Greeks only built paved roads to connect a city with a sanctuary. These roads had purposely build-in ruts—a groove or channel adapted to the ordinary span of a carriage for steadying and directing the course of the wheels, i.e., a "stone railway". The rut, termed ogmos, was the real "road" and the rest of the road was left unpaved. The gauge was 138–144 centimeters (7–10 cm deep and 20–22 cm wide).

The engineer of the Classical Period used five simple machines: the lever, wheel, the pulley, the wedge and the screw. These were combined into the sheerlegs—the earliest crane, constructed of two vertical spars, cross-piece, windlass, block and tackle. Greeks used iron in structures with great confidence. It was used as material for cantilevers and joinery in the Parthenon. Beams of wrought iron to transfer load to columns were used in column dovetail joints. Builders used limestone and marble, together with reinforcing iron members, in their beams. Greeks were familiar with mortar, having developed hydraulic cement. Lead and wrought iron were materials used for joinery. Greeks used post-and-beam construction in almost all their building. They understood the arch, but used it infrequently. Stonework was regularly coursed "ashlar" masonry. The Greek builder commonly used headers (bricks whose long axis lies through the wall, with only the end showing on the wall face) in courses for binding face-stones to the body of the wall. The Greeks, using only basic structural forms such as post-and-beam, refined this method to build some of the most beautiful buildings of all time.

The two classical building styles (Corinthian being a later Hellenistic and Roman Period style) were Doric and Ionic.

THE MINES OF ATHENS — LAURION

Typically, miners lived underground for a week and came out for a week. Abandoned stopes were utilized as kitchens and for sleeping quarters. Holes cut in the country rock were used for cupboards and lamp niches. Ores were lifted by small skips and hauled out by a rope, often guided over a wheel on the rim of a shaft.

Water was a limiting factor in the shaft depths. Pumping was not developed extensively until Roman times. Deep shafts such as those at Laurion (110 meters) necessitated air shafts (50 × 80 centimeters) for ventilation. Fires at the bottom of the shafts were probably utilized for draught. Ledges were cut for ladder rungs.

The geology at Laurion consisted of three beds of limestone or marble separated by two strata of mica-schists, the whole traversed by dikes of granite and gabbro. The silver-bearing galena was concentrated at junctions (tops) of lower beds of limestone. The galena, resting on zinc ore, was covered by iron ore. It was very rich in silver, containing 1,200–4,000 g Ag per ton of ore. Hematite was worked in the third contact zone. Shafts were sunk in pairs and parallel galleries driven from them. Frequent cross-cuts between galleries insured ventilation. One hundred thousand tons of ore was mined from such stopes. A miner averaged 4.5 meters per month. Timbering generally was not used in Laurion. The wealth they produced was to pass away in Athen's future disastrous war with Sparta (434–404 B.C.), when the site was prepared on the Athenian Acropolis for the crowning achievement of the building in the Classical Period.

THE PARTHENON(S)

The Parthenon was not the result of an overt collaboration between Kallikrates and Iktinus.[5] Hill (1912) convincingly presented evidence of two parthenons—an earlier, smaller version and the version seen today.[6] Kallikrates began the original Doric style structure around 465 B.C. during the reign of the Athenian ruler, Kimon. Under his charge until 449 B.C., the site was prepared and construction continued apace to the point that all exterior columns were mounted. The "Kimonian Parthenon" had six columns at either end and 16 along the sides. When Kimon died in 450 B.C. and his political adversary, Pericles, ascended to power, Iktinus took over as builder. This change seems less a judgment of Kallikrates' design than a patent move by Pericles to reward his own builder with the contract for the most prominent structure in Athens. This conclusion is supported by the fact that Kallikrates is credited with eight other major projects in Athens after 449 B.C., including the Hephasiteion (448–442 B.C.) and the crucial walls from Athens to its port at Piraeus.

Iktinus lengthened the Kimonian design a little to the west (67.36 to 69.49 meters) and widened it to the north (23.77 to 31.09 meters). The effect was to change the front by adding two columns and the sides by one column (6 × 16 to 8 × 17; cf. Figure 3-5). Iktinus' contribution to the final Parthenon lay not so much in the change in overall dimensions, but how these were integrated into the earlier Kimonian design to make it his own.

Proportions in Greek temples sprang from the given diameter of the columns. In the Parthenon, the ratio of the column base to the interaxial dimension was 4:9, which was repeated in the stylobate (colonnade substructure or platform), e.g., 31.09 × 69.49. To avoid the optical illusions of sagging in horizontal lines and bulging in vertical lines, Iktinus contoured the stylobate imperceptibly upward and canted the columns inward on a slight diagonal. In the stylobate, the rise is 0.127 meters and the declination of the columns is 0.076 off the vertical. Further, he used the technique of entasis in which the column is curved slightly convex to create the illusion of strict verticality. The column styles were mixed as well. The facades were Doric, as Iktinus probably used those already set in place by Kallikrates; the interior hall columns were Ionic. This was a first in Greek design and demonstrated the Ionic column's appropriateness to taller dimensions. Greek columns were built up of a series of drums stacked one atop another. In the center of each drum was a square hole or "empolia" for a wooden block which, in turn, was holed for a dowel pin. Seams created by this construction were barely noticeable.

The Greek temple, while representing little in the advance of engineering, was the height of an elegant style. The post-and-beam construction was a conservative building technique. Its use was seen in ancient Stonehenge—short spans of heavy stone horizontals in tension

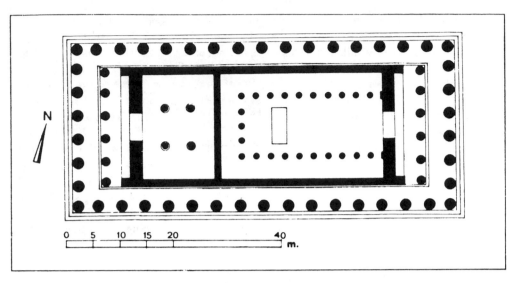

FIGURE 3-5.   Plan view of Parthenon and section of front portico. (Redrawn from Carpenter, 1970.)

held aloft by vertical columns in compression. The Greeks refined the aesthetic form, but dared little beyond it.

THE POST-CLASSIC-HELLENISTIC ENGINEERING

After the Athenian-Spartan conflicts of the 5th century, Greece drifted away from the cultural heights of the Periclean era. As proved by a fourth indecisive war, Sparta, without

Persian money, could not match Athens. Persia, itself, was having difficulties, with Egypt breaking away (ca. 400 B.C.), and the subsequent loss of her Indian provinces (ca. 380). In the western Mediterranean, Syracuse, the strongest of the Greek colonies, was matched against the rising city-state of Carthage. Syracuse could only confront her adversary in Sicily, while Carthage expanded westward across North Africa to the Iberian Peninsula.

By 352 B.C., a new force had arisen in the Greek political world. This was Macedonia. Phillip of Macedon (359–336 B.C.) transformed his "backwoods" army into the most efficient in Greece. By 338 B.C., he defeated the combined Theban-Athenian forces at Chaeronea and formed a pan-Hellenic league much like Athens a century before. Phillip's objective was really Persia, but this crusade was for his famous son, Alexander the Great. Alexander conquered Persia by 330 B.C., following his father's assassination in 336 B.C. The remainder of his short life (356–323 B.C.) was spent campaigning eastward into the Steppes and India. Upon his death, his empire splintered into sections ruled by his generals. While the tawdry story of division and conflict plays badly for the conquering Greeks, the economic and cultural spread of Hellenism more than redeems them.

The conquered territories embraced Greek culture in varying degrees—Anatolia became Hellenistic; Egypt, superficially so; Mesopotamia, little or none. The capitol of this pan-cultural expression was Alexandria. This new metropolis quickly outgrew its predecessors to over a quarter-million persons by the 2nd century. Greek learning, science and engineering thrived here. In the Hellenistic world, with Alexandria at its center, flourished perhaps the most gifted of all the Greek engineers. These included:

Aristotle (384–322 B.C.) or Straton. Modern day historians are uncertain whether to credit Aristotle or his student, Straton, with the first engineering treatise titled "Mechanics" or "Mechanical Problems", which discusses the lever, as well as gearing. Straton accepted Democritos' theory of atoms and discovered compression. *Mechanika* discusses all the simple machines except the screw.

Archimedes of Syracuse (287–212 B.C.). Archimedes discovered the principles of the screw and of buoyancy. In *Statics*, he treats the theory of the lever, the force parallelogram and calculation of pi by "exhaustion method", foreshadowing the infinitesimal calculus by 19 centuries.

Ctesibius of Alexandria (ca. 285–247 B.C.). Ctesibius invented the force pump, hydraulic pipe organ, metal spring, automated water clock and the keyboard.

Hero of Alexandria (ca. 100 B.C.). He understood the application of the vacuum and the bent siphon. He wrote treatises on mathematics and surveying in which he describes the dioptra (forerunner to the transit level) and pneumatics.

Philon of Byzantium (ca. 250 B.C.). Fascinated by the siphon, he designed automata using principle, wrote on mechanics. He was the first to describe the undershot waterwheel and appreciate its potential.

Euclid of Alexandria (ca. 300 B.C.). Euclid succeeded Thales and Pythagoras and was the greatest geometrician, followed closely by Archimedes. It was at Alexandria that Pythagoras' theorem was proven, although we do not know it if was done by Euclid himself.

PROOF OF PYTHAGOREAN THEOREM*

Consider a square of length a + b, where a = 4, b = 3 and c = 5.

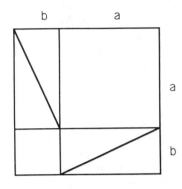 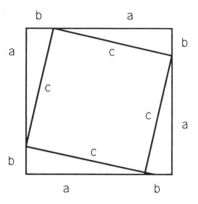

1. $A = 4ab + (b - a)^2$
    $= 4a + b^2 - 2ab + a^2$
    $= a^2 + b^2 + 2ab$
2. $A = c^2 + 4 \{1/2 \ ab\}$
    $= c^2 + 2ab$

    $A = A$
    $c^2 + 2ab = a^2 + b^2 + 2ab$
or $c^2 = a^2 + b^2$

Important Hellenistic engineering accomplishments include:

The Pharos of Alexandria. Built by Sostratos of Knodos in the 3rd century B.C., it was approximately 440 feet tall with light on upper tier.

The Colossus of Rhodes. Built by Chares of Lindos in 280 B.C., it stood for 56 years. Constructed to between 27 to 37 meters, it was sheathed in bronze on an iron armature supported on columns.

The Pergamon Siphon. A siphon was constructed at Pergamon during the reign of Eumenes II (200–190 B.C.), which carried water 56 kilometers in 7.62–17.78 centimeter pipes of tile side-by-side (Figure 3–6). The pipes discharged into a reservoir 3.2 kilometers from the city and 30 meters above it. From there, a single 25 centimeter wooden or bronze pipe, reinforced at joint sections by large perforated stone blocks, ran down across several valleys—the deepest point was 200 meters lower than the reservoir's elevation—and up over intervening ridges. At the low point in the line, water pressure approached 15 atmospheres within the pipe. Tile or lead pipes would have been unable to withstand this pressure.

Hero's interest in the bent siphon and, thus, the vacuum, led him to author *Pneumatica*, which became the starting point for the modern study of fluid dynamics.[7] His treatise was a study of "natural magic", or the unexplained properties of nature. He describes the first reaction turbine in his aeophile (Figure 3–7). In this simple device, Hero foreshadowed the work of later engineers in the use of steam as a motive force by 16 centuries.

* Proven in Alexandria; *not* by Pythagoras.

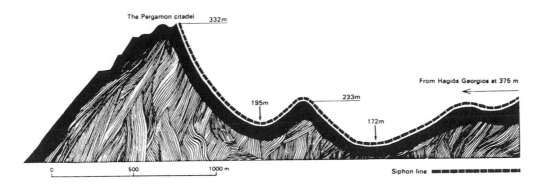

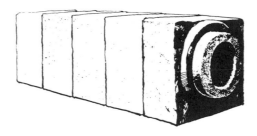

FIGURE 3-6.   Pergamon siphon with detail of supposed stone water pipe. (Redrawn from Sandström, 1970.)

FIGURE 3-7.   Hero's aeophile.

   While Hero mentions the force pump, it is Ctesibius who is credited by Plutarch with its invention (Figure 3–8, upper). His pump was constructed of bronze. The cylinders were oiled and the valving allowed water to be raised to a height consistent with the pump's ability to withstand the internal hydrostatic pressure. The valves were held in place by tempered iron springs, the first known use of this metal in such a fashion.[8]

   The other water-raising device from this period was the Archimedean screw (Figure 3–8, lower), named for this great engineer. This machine, also known as the water-snail, was described by Vitruvius[9] in the first century B.C., with clear instructions for its manufacture. The rotor consisted of a wooden cylinder with a length/diameter ratio of 16:1. At either end, the circumference was divided into eight equal arcs. The length was divided into sections, each equal to one-eighth of the circumference. The blade was drawn obliquely around the rotor shaft one-eighth of the circumference along and at the same distance around

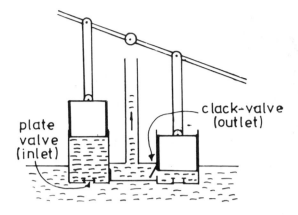

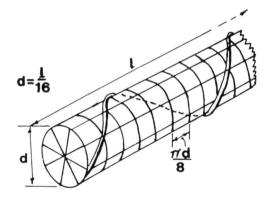

FIGURE 3-8.   *Upper:* The force pump. *Lower:* The Archimedean screw. (Redrawn from Hill, 1984.)

the shaft with the resultant pitch of 45° (see Figure 3-8, lower). The winding was continued in this way until a laminated blade had been built up to twice the diameter of the shaft. Seven additional blades were formed in the same way. A wooden case fit around the blades. Of barrel-like construction, the planks were painted with pitch and bound with iron hoops. The description of the bearings is somewhat obscure, but iron caps were probably provided for the ends of the rotor from which iron spigots protruded. The spigots rotated in iron journals.

A final device of these early mechanical engineers was a fascinating mechanism found in a shipwreck near the island of Antikythera, between Crete and Kythera. The Antikythera Mechanism represents a complex analog device for the calculation of time and astronomical cycles. It was probably built around 87 B.C. in Rhodes.[10] The device consisted of finely turned bronze gear wheels. The reconstruction shown in Figure 3–9 illustrates an analog computer of great sophistication. Until this discovery in 1900, the level of fine technology in Hellenistic times was not believed to be this high. We do not know to what extent this skill was used, only that the knowledge existed at both theoretical and practical levels.

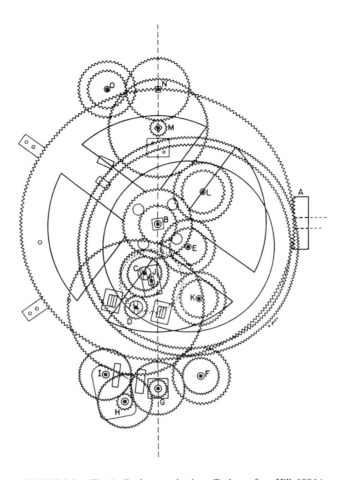

FIGURE 3-9.   The Antikythera mechanism. (Redrawn from Hill, 1984.)

NOTES

1. **McEvedy, C.** *The Penguin Atlas of Ancient History*. Penguin (1967).
2. Ibid.
3. **Kirby, R. S., Thington, S. W., Darling, A. D. and Kilgour, F. G.** *Engineering in History*. McGraw-Hill (1956).
4. **Wood, M.** *In Search of the Trojan War*. Facts on File Video (1985).
5. **Carpenter, R.** *The Architects of the Parthenon*. Penguin (1970).
6. **Hill, B. H.** The older Parthenon. *Am. J. Archaeol.* **XVI**: 535–58 (1912).
7. **Boas, M.** Hero's *Pneumatica*, a study of its transmission and influence. Seminar in the History of Science, Cornell (1960).
8. **DeCamp, L. S.** *The Ancient Engineers*. Ballantine (1963), pp. 142–143.
9. **Vitruvius, M. P.** *De architectura libri decem.*
10. **Hill, D.** *A History of Engineering in Classical and Medieval Times*. Open Court (1984), p. 186.

4 ROME — THE ENGINEERING OF WAR AND CITIES

The name for Rome comes from the Latin term for river (''fiume''), which in dialect is ''ruma''. Roman history consists mainly of two periods: (1) the Republic (ca. 509–27 B.C.) and (2) the Empire (27 B.C.–476 A.D.). The Republican period was the ''Age of Conquest'', while the first 200 years of the Empire was ''Pax Romana'', or the ''Roman Peace''. The age of the Republic was the age of the military engineer; the age of the Empire, that of the civil engineer.

Early Rome was a village of farmers surrounded and governed by Etruscan kings. Rome threw off Etruscan rule with the overthrow of Tarquin the Proud in 509 B.C. Rome was sandwiched between the Etruscans on the north and the Greek colonies on the south (Figure 4-1). The early Romans were farmers. Their society was patriarchal, one with a pantheon of ancestor gods analogous to that of the Greeks, e.g., Zeus:Jupiter; Ares:Mars; Athena:Minerva: Venus:Aphrodite; and Apollo:Mercury. Believers in soothsayers, omens and animism, Romans, in short, were superstitious. In Republican times, Rome was an oligarchy. Patricians ruled through the Senate with elected consuls (2) who served annual terms. In periods of emergency, a dictator could be appointed for the duration (Cincinnatus was one of the first famous dictators).

Rome was founded on the Tiber River on seven hills. Legend has the area of the city the distance of the circumference the first ''king'', Romulus, could plow in a day. The Romans learned much of their engineering from other peoples. They probably inherited the arch from the Etruscans. From the Greeks, they learned about water supplies by means of aqueducts, techniques of tunnelling and the use of hydraulic cement. Road building was learned from either the Etruscans or the Greeks, but Rome elaborated this skill to a level unmatched until modern times. The Romans have been called the greatest engineers of ancient times. The key to their success lay in their ability to organize and administer large-scale projects and their practical approach to engineering problems. They did not, however, contribute many fundamentally new ideas or original concepts. Their achievements, in sheer size and the scope of construction, were remarkable, as was their determination to bring them to fruition.

Rome was, first and always, a military power. She owed her long history to her success on the battlefield. Rome exhibited a dogged tenacity that overcame disastrous reversals against, first, the Celtic tribes, Carthage and, later, the Parthians and Germans. Persistence was a Roman virtue. In their society, planning and predictability took precedence over precocity. Roman military commanders were more apt to lose by adhering to a set order of battle rather than to win by innovation. That which was successful in the past was not easily cast aside by the Roman mind.

Such a description of the Romans does not exclude the ease with which they adopted things that benefited them. Marcus Porcius Cato Censorinus, though an enemy of Greek culture, erected the first basilica in the Forum in 184 B.C.[1] In the Punic Wars, Roman landsmen became sailors to defeat the traditional naval power of Carthage. It has been said in various forms that ''Rome conquered Greece and, in turn, was conquered by Greece.'' It is obvious that Hellenistic design and principles did become the foundation of Roman engineering.

Again, it was in relation to Roman military action that this is most aptly demonstrated. Grecian Syracuse came down on the wrong side (from the Roman view) of the dispute between Rome and Carthage during the Second Punic War (218–201 B.C.). Archimedes

Italy

FIGURE 4-1. Early Rome.

created a formidable array of defenses for Syracuse.[2] His catapults and cranes disrupted the Roman assault to the point that the commander, Marcellus, decided on a drawn-out siege instead.[3] With the fall of Syracuse in 212 B.C., Archimedes was killed by an over-zealous legionary after Marcellus had given orders to capture the great Greek engineer alive. As a historical vignette, we can only guess at the accuracy of the circumstances surrounding Archimedes' death. Nonetheless, it expresses the respect which was held by Rome toward Grecian works of engineering.

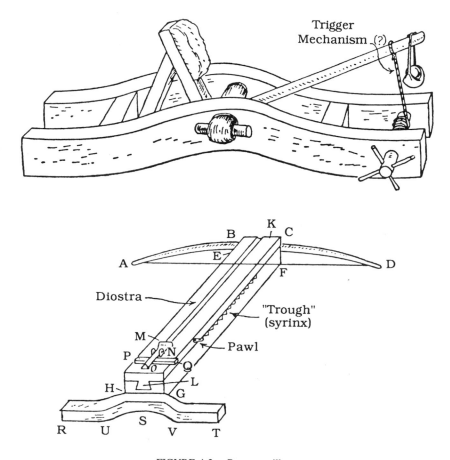

FIGURE 4-2. Roman artillery.

The early army of Republican Rome was a volunteer army that was formed out of voting (centuries) political units: when the campaign was over, the army stood down. Generals were *consuls*, with *tribunes* (2) in charge of the legion (*legio*). The Republican legion was composed of centuries organized into maniples (3). The maniples formed into three battle lines which could reinforce each other and form a solid phalanx or line of heavy infantry with spears. Cavalry was used sparingly. Only Roman citizens (landowners) could serve in the legions. The army remained so structured until the end of the Punic Wars (265–241 B.C.; 218–201 B.C.; 149–146 B.C.) in which Rome wrested control of the western Mediterranean from Carthage. In the First Punic War, Rome obtained Sicily; the Second Punic War was the era of Hannibal; and the Third Punic War saw the destruction of Carthage by Rome out of fear and spite.

Rome added Macedonia and Greece in 146 B.C. The Republic began to collapse with civil strife and social revolution (133–121 B.C). Marius, founder of the new professional army, was replaced by the autocrat Sulla. Marius created an army of landless urban poor whose allegiance was to the general, not the Senate. This army, based on the cohort (500 men) of centuries (100 men), formed into legions. The legate (*legatus*) replaced the tribunes and the centurions were tactical leaders.

Important to the new army was the military engineer. Military engineers were responsible for siege engines and artillery. They also built roads, bridges, baths and aqueducts. Mechanized arms were built of wood. The onager (mule) (Figure 4–2) was a great sling which used stone cannonballs of 1 to 5 kilograms, attaining a range of roughly 0.5 kilometer. The wooden arm was bent back, under pressure of catgut strings, to a catch. When released,

TABLE 4-1
CALCULATION OF THE CALIBER OF CATAPULT

Multiply by 100 the weight (in "Minas") of stone projectile, then

$$d \text{ (in fingerlengths)} = 1.1 \times \sqrt[3]{100} \times \text{Minas}$$

Example: If stone = 80 Minas (35 kg)
$$80 \times 100 = 8,000$$
$$\sqrt[3]{8,000} = 20$$
$$0.1 \times 20 = 2$$
$$d = 20 + 2 \text{ or } 22 \text{ fingerlengths caliber } (\approx 110 \text{ cm})$$

Round off fractions by addition of 1/10

the force developed was 50,000 kg cm^{-2}. Catapults were mechanized bows (Figure 4–2). A modern reconstruction in Saalburg Museum in Germany shot a large arrow, 289 meters, against the wind. Arrows of 1 meter went 335 meters; lead balls (0.5 kg) went 274 meters. Arrow penetration was least 3.8 centimeters when striking a shield of thin iron plate 3 centimeters thick. The ballista fired projectiles from a beam mounted on a platform. The arms of the ballista worked against the tension of wound strands when winched backward (cf. Table 4–1 for caliber calculation).

Pompey conquered Asia (66–62 B.C.); Caesar conquered Gaul (58–51 B.C.) and made two expeditions to Britain. Caesar returned to Rome in 49 B.C., defeated Pompey and was assassinated in 44 B.C. The Triumverate ruled until Octavian became the first emperor in 31 B.C. (31 B.C.–14 A.D.).

From 27 B.C. to 180 A.D., Pax Romana existed with a standing Roman army at the frontiers (Figure 4.1). Veterans of 16- to 20-year enlistments set up colonies in new provinces. A small, permanent navy was formed to guard against piracy. Taxes, at 10% of earnings, were instituted. After Nerva (96–98 A.D.), successors were "adopted" for ability rather than hereditary succession.[4] In the 3rd century, the Germanic tribes were a problem and Asian wars became more frequent, particularly with Parthia. By the time of Constantine the Great (324–337 A.D.), Christianity had become the state religion. Constantinople became the capitol of the Eastern Roman Empire and Rome, the capitol of the declining West. Attila and the Huns (372–451 A.D.) repeatedly threatened the Western Empire until defeated in 451 A.D. at the Battle of Troyes by Aetius, a last great Roman commander. The fall of the Western Empire occurred in 476 A.D. The Eastern Empire survived under emperors at Constantinople until 1453 A.D.

Rivoira[5] states that, by the time of Augustus, most Roman "architects" came from the ranks of military engineers. Within the army of the 2nd century A.D., the following specialists were attached to a legion (1 legion = 6,000 men):

1. architectus — "master builder"; rank: centurion or tribune
2. mensor — surveyor; rank unknown
3. hydraularius — water engineer; rank: centurion or tribune
4. ballistarius — catapult/artillery maker; rank unknown

The Romans fortified their camps when campaigning. These defensive sites could either be only temporary "marching camps" or evolve into legionary fortresses. After successful campaigns, "Romanization" began apace with settlements ofttimes or "coloniae" founded. In Gaul, tribes were resettled to sites that grew into towns, later becoming smaller urban copies of Rome itself. The evolution of fortifications is illustrated in Figure 4–3.

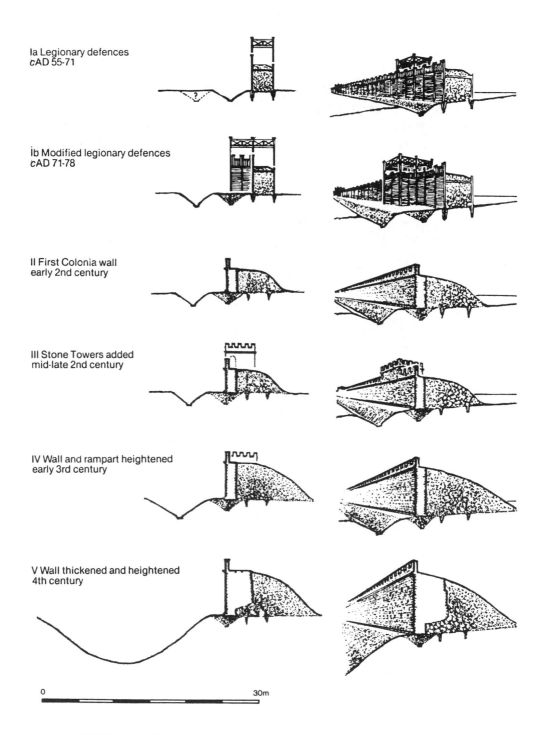

Ia Legionary defences
cAD 55-71

Ib Modified legionary defences
cAD 71-78

II First Colonia wall
early 2nd century

III Stone Towers added
mid-late 2nd century

IV Wall and rampart heightened
early 3rd century

V Wall thickened and heightened
4th century

0   30m

FIGURE 4-3. The evolution of Roman fortifications at Lincoln. (After Jones, 1983.)

ROADS

Rome built 300,000 kilometers of roads; 90,000 kilometers of "paved" viae (Figure 4-4). The armies could travel on highways from Britain to the Euphrates (4,000 km). They made possible trade and communication and postal service. The origin of the Roman road has been rightly suggested as Etruscan in nature. Etruria ruled Rome until the 5th century B.C. and it is logical that these people taught much to the more agrarian Romans, particularly in areas such as bridge and road building. The Romans classified their roads according to use. In 450 B.C., the Twelve Tables (oldest Roman legal code) made the following classification:

1.   semita — foot path, 0.3 meter wide
2.   iter — for horsemen and pedestrians, 1 meter
3.   actus — single carriage, 1.2 meters
4.   via — two-lane, 2.5 meters

In later centuries, these dimensions lost meaning as via widths increased to over 7 meters. Typical thickness for viae were 1 to 2 meters.

Two primary sources on Roman road construction are Statius, a Roman poet, and Vitruvius, an engineer, both writing in the 1st century A.D.

Here is what Statius tells us about the building of a great highway (via) under the emperor Domitian (about A.D. 90):

> Now the first state of the work was to dig ditches and to run a trench in the soil between them. Then this empty ditch was filled up with the foundation courses and a watertight layer or binder and a foundation was prepared to carry the pavement. For the surface should not vibrate, otherwise the base is unreliable or the bed in which the stones are rammed is too loose. Finally the pavement should be fastened by pointed blocks and held at regular distances by wedges. Many hands work outside the road itself. Here trees are cut down and the slopes of hills are bared; there the pickaxe levels the rock or creates a log from a tree; there clamps are driven into the rocks and walls are woven from slaked lime and grey tufa. Hand-driven pumps drain the pools formed by underground water and brooks are turned from their courses.

Principal Roman roads (Figure 4-5) typically consisted of four layers, with individual levels as follows:

statumen: 0.5 meter on earth subgrade leveled with sand
rudus: 22 centimeters stone with mortar
nucleus: 22 centimeters fine-gravel concrete
pavimentum: hard stone blocks up to 1 meter square

Curbs were 44 centimeters high with side paths 0.5 to 2.5 meters wide along the *viae*. These are more thoroughly described by Vitruvius:[36]

> First I will begin with rubble paving, which is the first stage in finishing, so that account may be taken with special care and great foresight, of a solid foundation (*statumen*). If we must carry our paving on level ground we must inquire whether the soil is throughout; it is then to be leveled, and rubble must be spread over the surface. But if there is a made site, in whole or in part, it must be rammed carefully with piles.

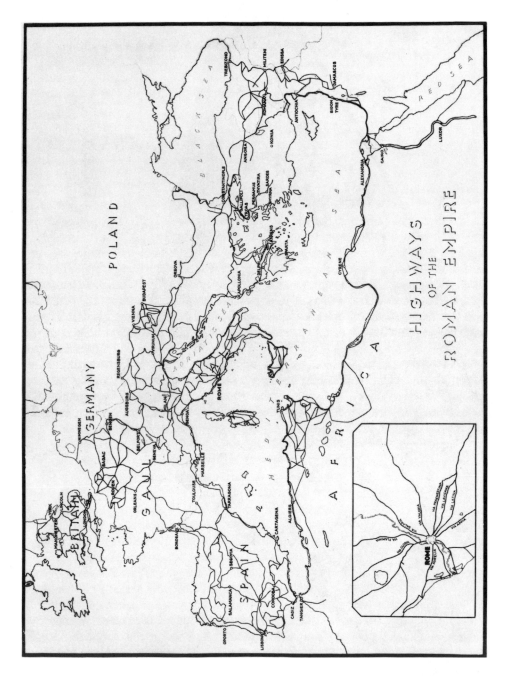

FIGURE 4-4. Farthest extent of Roman road system under Trajan (U.S. Bureau of Roads).

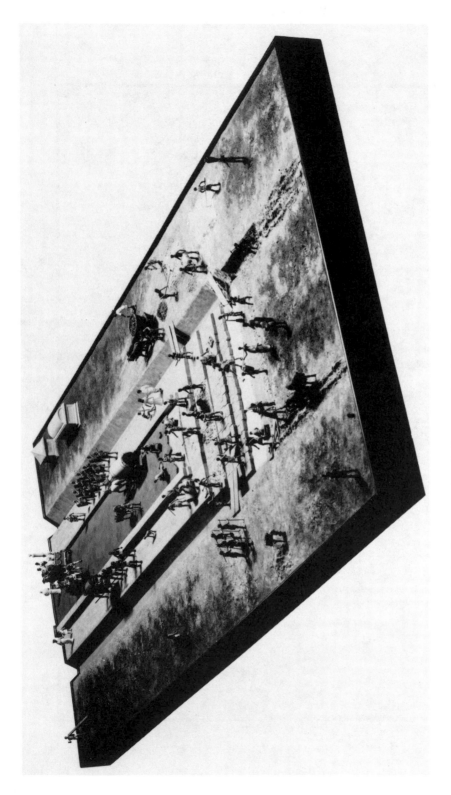

FIGURE 4-5.   Idealized construction of a Roman paved road—The Appian Way (U.S. Bureau of Roads).

He then goes on to say:

> A layer of stones (*rudus*) is to be spread each of which is not less than a handful. After spreading the stones, the rubble, if fresh, is to be mixed, three parts to one of lime; if it is of old materials five parts of rubble are to be mixed with two of lime. Let it then be laid on, and rammed down with repeated blows by gangs of men using wooden stamps. When the stamping is finished it must be no less than nine inches thick, that is, three-quarters of its initial height. On top of this (the *rudus*) another layer is laid (*nucleus*) consisting of a hard coat of powdered pottery, three parts to one of lime, forming a layer of six inches. On the finishing coat a surface layer (*pavimentum*) is laid.

This description is also that of a Roman house floor. We know from excavations that it applies to Roman roads. The statumen and rudus together can be compared with our hand-laid foundations in which the rudus is a damp course shutting out water from the statumen. The nucleus is a kind of binder course for the pavimentum proper.

The Roman engineer typically had very good materials at his disposal. He adopted local materials insofar as possible, locating building quarries where he built his road. This resulted in the use of widely diverse materials, as well as types of construction. Lime mortar was adopted following Hellenistic practice (ca. 300 B.C.). The Romans used it in combination with rubble, potsherds and crushed bricks to prepare a grout which could penetrate between the stones of the foundation layers, thus giving the solidity and resemblance of a wall. The Roman road has been characterized as a "wall on the flat". Thickness for the main highways varies from 1 to 2 meters, more than four times the thickness of the best concrete roads of today. Their average life, which we can determine from the inscriptions and milestones, was about 70 to 100 years. It has been suggested that Roman structures survived the passage of time because the mass of their construction was out of all proportion to function. Their roads were for the legions and were too steep in places for wheeled traffic. Maintenance was an expensive problem as well. Still, the Roman attitude was to build for longevity. Considering the traffic of vehicles and unshod animals, both of which were fairly hard on a road surface, this was a good idea.[6]

The grouting of gravel surfaces resulted in a concrete road. The invention of true hydraulic concrete and its application in architecture and civil engineering was the only great discovery that can be ascribed to the Romans. About 150 B.C. near Puzzeoli, they discovered the natural strata of trass (porous volcanic rock), a valuable substitute for lime which came to be known as pozzolana.[7]

As the Roman roads were built, the engineers often came upon subsurface soil conditions not suitable normally for highway use. One special method devised for crossing marshy ground is still in use in remote locations in the United States and Canada, where the construction method is referred to as a "corduroy road".[8]

> Two rows of cross-beams were placed on the marshy soil for the whole length of the road. The distance between the beams was 2m. There was a gap between the internal ends of each pair of beams, while the external ends projected some 40 cm beyond the basic timber seating. These outer ends were slotted to take vertical stakes, so pinning the framework to the ground. On the cross-beams were fixed lengthwise with the road two lines of joists to carry the sides of the highway. The joists bore a transverse "corduroy" of tree trunks and on this, in turn, there lay limestone flags cemented with clay, covered again by the road-metalling of gravel and pebbles. Tree trunks were discovered under other roads at river crossings.[9]

Roads in England, due probably to climate, had a 12 centimeter bottom layer of cobbles, 45 centimeters of soft limestone and earth, 45 centimeters of rammed earth, statumen of

FIGURE 4-6.   Italian peninsula and route of Appian Way. (Redrawn from Kirby et al., 1956.)

boulders and broken stone. Another variant in London (*Londinium*) was 2.2 meters of gravel concrete between retaining walls with white clay tile surface paving. Watling Street ran from Dover to Canterbury, London to Chester, then to Lancaster and beyond.

Roman roads, on long traverses, do not deviate more than 0.4 to 0.8 kilometer in 30 to 50 kilometers. Grades of 20% (1 m in 5) were common, much steeper then today. Road cuts could reach 36 meters, which is significant even by today's standards.

The first major road, the *Via Appia*, was built by Appius Claudius Crassus ca. 310 B.C. Its route is shown in Figure 4-6. It was built in response to the Samnite Wars of the 4th century and opened up the Campania to Roman influence.[10] Eighty years later, Gaius Flaminius had built the *Via Flaminia*, the first great northern route (Figure 4-6). Flaminius

FIGURE 4-7.   Extract from the *Peutinger Table*. (Vienna State Library)

later fell in battle against Hannibal during the Second Punic War, at Lake Trasimenus (207 B.C.).[11]

Milestones were placed along the sides of the Roman roads with inscriptions indicating distances to one or more important places, the termini of the road, the road's identity, the names of the road builder and the emperor in reign at the time, and the name of the town or district which placed the milestones. Distances were expressed in thousand-pace increments, denoted as M.P. in the inscriptions. The milestones were cylindrical columns three to nine feet (Roman) in length, commonly six feet (Roman) long, and standing on a plain base with a cylindrical neckmould.

Later, Golden Milestones were established in major cities, with listings of distances to all major destinations on all major roads leading from the city. While not true milestones, they were, in effect, the zero milestone for the city, listing the itinerary of destinations. The most well-known road map or guide is the Peutinger Table, now displayed in the Library of Vienna. It is 5.5 meters long and 30 centimeters wide, a 13th century copy of the Roman original (Figure 4-7). The map outlines all parts of the known world at the time of Theodosius (378–395 A.D.), with the exception of parts of the map relating to Spain and Britain which are lost. The Peutinger Table (named for Conrad Peutinger of Augsburg)[12] has been ascribed to Castorius, a Roman cosmographer ca. 365 A.D.[13] The map has little geographical accuracy and was, at best, an itinerary for decision-making by the user.[14]

SURVEYING

Mensuration was developed to a science by the Romans. The *agrimensor* (surveyor) used either the *groma* or *dioptra* (Figure 4-8). Romans preferred the *groma*, a cross-tie arrangement of plumb lines suspended from ends (Figure 4-8b). The *dioptra*, ancestor to the transit, consisted of a sighting bar free to swing through either a horizontal or vertical arc (Figure 4-8a). The chaining "rod" was 10 feet (Roman) and called a *decapeda*. For leveling, *chorobates* (20 Roman feet) was straight-edged with two equal, right-angle legs. A groove with water provided supplemental level. The *libella* was used for leveling shorter distances. For distance measure, a preambulator or odometer of known circumference was also used.

The principal literary source on Roman surveying instruments and practice is the *Corpus Agimensorum*. Strangely enough, the Corpus does not give an illustration of the groma.[15] It is from archaeology that we have obtained an example of a groma. In 1912, excavations at Pompeii discovered the workshop of a surveyor called Verus. The reconstruction in Figure 4-8b was based on this find. In later Latin times, the surveyors called themselves *gromatici* after this most usual instrument of their profession.[16]

Roman surveyors laid out roads, but their principal task lay in centuriation or the dividing up of land.[17] Centuriation, as the name implies, used as its standard measure an area roughly 500,000 square meters (2,400 Roman feet or 20 actus per side). In Roman measure, the century was 200 *ingera*. Typically, land was centuriated from a point defined by the intersection of two principal axes (in most instances, these were roads). These axes were termed the *kardo maximus* (K.M.) and the *decumanus maximus* (D.M.). The *kardo* was oriented north-south and the *decumanus*, east-west.

Land units, or in towns, blocks (*insulae*), were keyed locationally to the two main roads. Positions were referenced as (1) left (S.D.) or right (D.D.) of the decumanus and (2) near (C.K.) or beyond (V.K.) the kardo.[18] Calculation of area was based on the ingera and given in Columella's *Agriculture*,[19] in which field shapes were approximated by geometrical figures such as the square, trapezium and segments of a circle. Military surveyors measured distance in miles (*milia*) and paces (*passus*) or miles and feet (*pedes*). Surveying also was critical to the proper run of an aqueduct. Here leveling was as important as the line of traverse.

AQUEDUCTS AND WATER SUPPLIES

In Rome alone, there were 11 aqueducts which supplied 1,168,850 cubic meters (113 $\times 10^6$ liters or 570 liters per capita). The estimates for the various aqueducts of the city are shown in Table 4-2.[20]

Sextus Julius Frontinus was water commissioner ("Aqualegus") for the city of Rome. In 97 A.D., he wrote a two-volume work on the water supply for which he was administrator.[21] Table 4.2 of 11 aqueducts represents the water supply to the city at its height during the second century A.D., with a population estimated between one million and two million persons. Frontinus himself only oversaw nine aqueducts, as Trajana (109 A.D.) and Alexandrina (226 A.D.) were added to the system later. The four principal aqueducts were Anio Vetus (272 B.C.) and Novus (38–52 A.D.), Marcia (144–140 B.C.), and Claudia (38–52 A.D.).[22]

Roman aqueducts were based on a low-pressure, continuous, gravitational flow theory of hydraulics.[23] Hero, probably contemporary with Vitruvius and Frontinus, had clearly demonstrated that flow (Q) is equal to the cross-sectional area (A) and flow velocity (V) or

$$Q = A \times V$$

He had conducted a novel experiment by damming a small spring and observing its discharge through a measured outlet.[24] To Vitruvius and Frontinus, their conception is best described as

$$Q = A$$

with no consideration of flow velocity or pressure head.

Technical divisions within Roman aqueduct engineering were: (1) agrimensors (land surveyors); (2) librators (levelers); (3) mensors (quantity measurers); and (4) aqualegus (aqueduct inspector). The earliest Roman engineer whose name is attached to an important aqueduct is Appius Claudius, builder of the Aqua Appia in 312 B.C. The Aqua Appia provided a low-level supply, largely by tunneling. It was the first of Rome's 11 municipal aqueducts (Figure 4-9).

The aqueduct Anio Vetus was constructed in 272 B.C., followed by the Aqua Marcia in 144 B.C. This latter aqueduct surmounted the problem of achieving high-level supply across the relatively flat Campagna, which surrounds the higher elevation of the city built on its famed "seven hills". Due to the lack of pressure pipes, water conduits had to be carried at grade. With the Aqua Marcia, advantage was taken of the highest supporting ground, coupled with the shortest distance across the Campagna (Figure 4-10a). A long cut-and-cover line had to be built circling the high ground of the Anio River and thence across the Campagna on masonry arches. The Marcia's length from intake at the Anio to the city was 93 kilometers. Ten kilometers of this length was on the masonry arches to bring water at 68 meters above the Tiber River level. The Marcia's waterway (specus) was 1.4 meters high and 2.8 meters wide. In 125 B.C., a smaller waterway, the Aqua Tepula, was built atop the Marcia. This was followed by the Aqua Julia (piggy-back on top of the Tepula) in the next century (Figure 4-10b).

In other areas of the Empire, great aqueducts were constructed, particularly in the province of Gaul (modern day France). The greatest of these include the Pont du Gard, 45 meters high and 275 meters long, set with massive piers and arches, and part of a 38-kilometer aqueduct bringing water to the Roman city of Nemausus (modern Nîmes). Its specus was 1.4 meters wide by 1.7 meters high. Concrete was used only in the *specus*, or waterway, with the piers being "dry" masonry.

Most aqueducts were built in the cheaper materials of brick and concrete. The construction of the Aqua Claudia (52 A.D.) (Figure 4-10a) marked a temporary return to the masonry style of construction. Stone arches carried the Aqua Claudia for 15 kilometers. The Aqua Anio Novus was completed atop the Claudia.

Based on data supplied by the aqualegus, Frontinus, a constant supply of 144 liters per capita was available in 52 A.D. Other writers put a maximum supply value of 450 liters per capita divided into usage percentages of: 17% industrial, 39% private, and 44% military (primarily barracks).

For all of their engineering achievements, the Romans never understood the mechanics of water flow. For example, Frontinus simply checked the area of the aqueducts against that of the outlets in the city. The Roman unit of flow, the *quinaria*, neglected flow velocity.[25] Frontinus did observe that the differences in head (pressure) made some pipes "take up more than their due measure."

At the same time of the building of the Aqua Claudia, Roman engineers undertook to relieve flooding of rich agricultural lands bordering Lake Fucinus. The project involved a tunnel 5.6 kilometers long under Mount Salviano and the Palantine Plains to an outlet into the Liris River. Two sections under the Plains were at the depth of 114 meters. Thirty thousand men were employed on the project for 11 years.

In Rome, the purity of water in aqueducts was maintained by covered channels, reservoirs (*castellum*) and settling basins much like the Greeks. There were strict laws against pollution

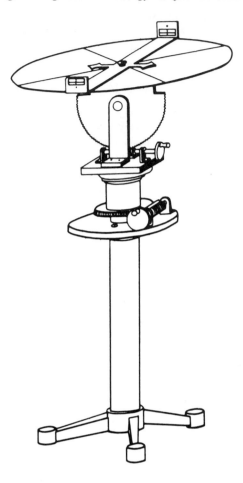

FIGURE 4-8a.   Dioptra. (Courtesy of Professor O. A. W. Dilke.)

and water theft. Vitruvius warned against lead pipes, but this went unheeded. Some aqueducts carried water unfit for drinking and were used for other purposes. Romans recognized the temperature and taste of water from various aqueducts and had definite preferences.

Water from the aqueducts was distributed to public fountains, industries (textiles), some private houses and public baths. The baths used prodigious amounts of water. The reservoir of the baths of Caracalla contained 76,124 cubic meters or 76,124,160 liters. Furnaces, where hot springs were not available, raised water to the desired temperatures and heated the floors and walls of the baths (hypocausts) (Figure 4-11). Excess water from the aqueducts was used to remove sewage via the Cloaca Maxima. This sewer system doubled as a storm drain system. Public latrines were maintained in Rome.

The majority of Romans lived in insulae, the large apartment houses which were often too tall for their foundations and supports. Augustus set a maximum height limit of 24 meters for private buildings. No running water was inside these insulae. Public latrines (most used interior commodes) were available in some insulae.

The great sewer of Rome, the Cloaca Maxima, was 5 meters high and, in places, 4 meters wide. There was no sewage treatment, with waste being discharged into the Tiber River and thence to the Mediterranean Sea. Floods backed up into these drains.

With the fall of the Western Empire, a lack of central authority and large public funds led to a decline in public services such as road building and water supply.

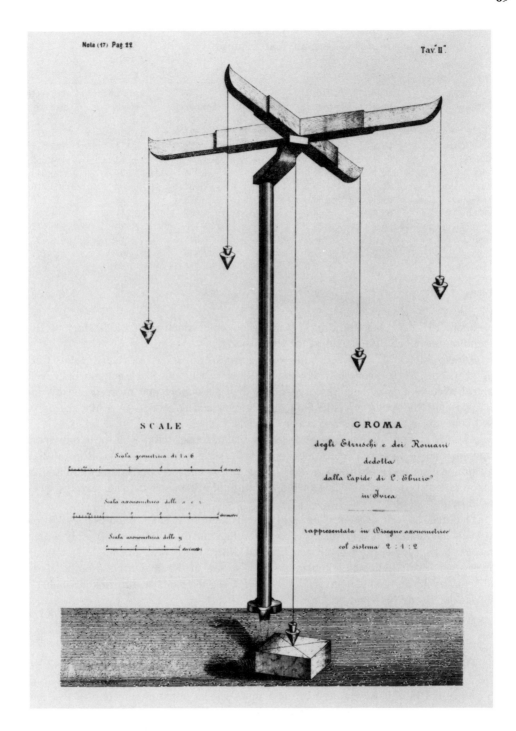

FIGURE 4-8b.   Groma (U.S. Bureau of Roads).

BRIDGES AND BUILDINGS

Marcus Vipanius Agrippa (63–12 B.C.) was the builder of the Pont du Gard, a monumental bridge and aqueduct that brought water to the Gallo-Roman city of Nemausus (modern-day Nîmes, France).[26] He erected the first public baths in Rome and the original

TABLE 4-2
Status of Roman Water Supply circa A.D. 300

Aqueduct	Length, kilometers	Head, meters	Gradient	Capacity		
				Quinaria	Liters per second	Cubic meters per day
Appia	16.6	8.0	0.5	1,825	876	75,737
Anio Vetus	63.9	219.7	3.4	4,398	2,111	182,517
Marcia	91.4	270.3	3.4	4,690	2,251	194,365
Julia	23.1	279.2	12.0	1,206	579	50,043
Virgo	21.2	5.3	0.3	2,504	1,202	103,916
Alisietina	33.3	136.0	4.1	392	188	16,228
Claudia	68.8	248.2	3.6	4,607	2,211	191,190
Anio Novus	86.9	428.2	4.9	4,738	2,274	196,627
Tepula	18.0	136.0	7.6	445	214	18,467
Trajana	57.0	72.0	1.3	2,846	1,367	118,127
Alexandria	22.0	9.0	0.4	521	250	21,633
Total	**502.2**			**28,172**	**13,523**	**1,168,850**

Pantheon.[27] Bridges and Rome have been synonymous since the famous story of Horatio defending the Tiber Bridge against the Etruscans.[28]

Roman armies on campaign built wooden truss bridges across rivers. Julius Caesar's engineers crossed the Saône in 1 day and the Rhine in 10.[29] As Roman engineers developed control over their pozzulana-based concrete, they built more and more using this media. Agrippa and other engineers still built their bridges and aqueducts in ashlar masonry, but used concrete in the footings and water channels.[30]

Another reason the Romans persisted in ashlar construction was the requirement for large amounts of wooden centering or forms for casting concrete. In this, Roman economy and pragmatism are probably best demonstrated. The censors, consuls and, subsequently, emperors lavished large amounts of monies on public works, but made savings wherever possible.

The arch, as well, lent itself to masonry construction. It is the main point of difference between the Romans and Greeks.[31] The beautiful Pont du Gard is really only a stack of arcades. Effective but expensive, later bridge builders used either the single arch design or lower spans of multiple semicircular arches. The use of the semicircular arch determines the span, as the arch must be exactly half as high as the span's width. Stated another way, the ratio of rise to half-span is 1.0.

When the arch is less than unity, it is termed segmental.[32] The Pons Fabricus (62 B.C.) has a rise/half-span ratio of 0.83. The spans of this bridge equal 24 meters. The longest spans in Roman bridges approached 42 meters.

The bridge across the Danube near the Iron Gate was the most pretentious of all Roman bridges. It was built in 104 A.D. by Apollodorus of Damascus who, along with Agrippa, Vitruvius and Frontinus, was one of the Empire's most noted engineers. All that remains of the bridge is to be found on the Trajan's column in Rome, together with a few references by contemporary historians and archaeological accounts of the pier foundations which may still be seen at low water. The bridge was almost 1,000 meters long and 15 meters wide.

Their hydraulic concrete, together with wooden pile footings, allowed the construction of piers that still stand today. Cofferdams, or dewatering structures, allowed the Romans to build in the water. Here they used archimedean and force-type pumps to remove the water from the cofferdam interior.

Concrete allowed the Roman engineers to vault enormous interior spans such as that of the Pantheon (Figure 4–12). Middleton[33] points out the rigidity and reduced lateral thrust

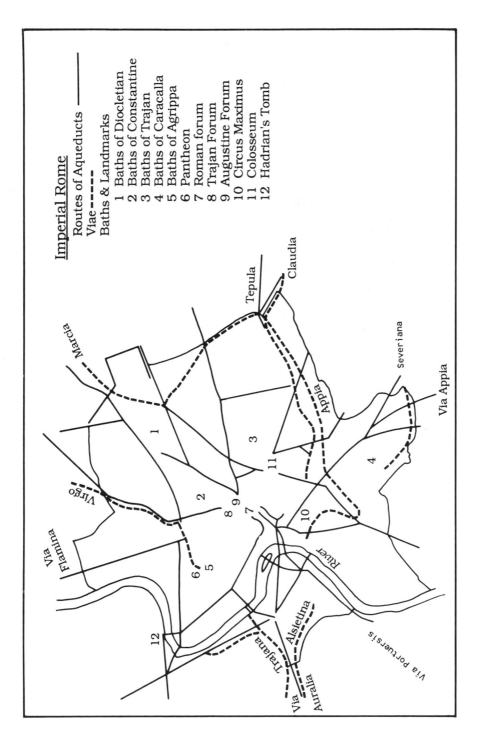

FIGURE 4-9. Routes of major aqueducts and baths of Rome. (Redrawn from Hammond, 1984.)

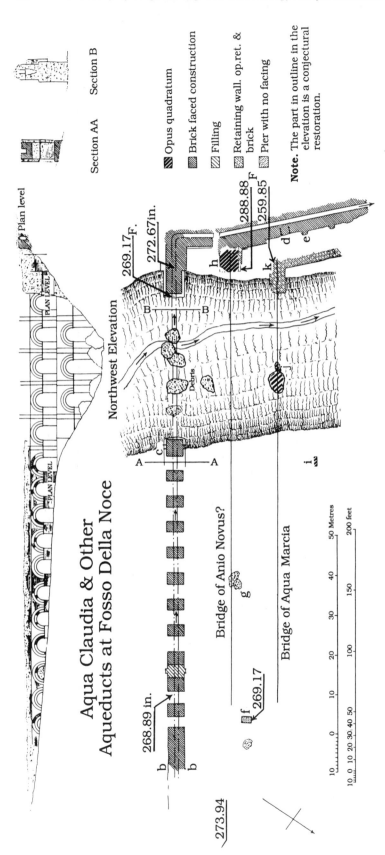

FIGURE 4-10a.   Aqua Claudia and other Aqueducts at Fosso Della Noce.

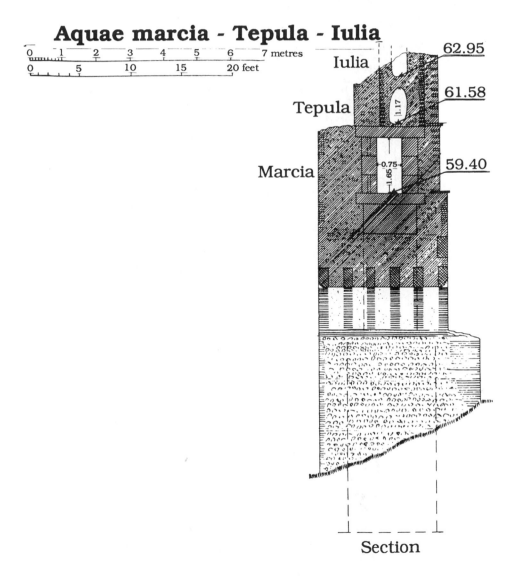

Aquae marcia - Tepula - Iulia

Iulia

Tepula

Marcia

62.95

61.58

59.40

1.17

0.75
1.65

0 1 2 3 4 5 6 7 metres
0 5 10 15 20 feet

Section

FIGURE 4-10b. Aquae Marcia-Tepula-Julia.

of the concrete vault. This assumption has been questioned by Robert Mark.[34] In his structural analysis of the Pantheon dome, he determined that the concrete was cracked into pie-shaped segments. These segments acted separately from one another and behaved as arches. To reduce tension caused by wall deflection, the step rings act as extrados used in the semicircular arch. The Pantheon was first built by Agrippa. His design consisted of an oblong cell and a large circular plaza.[35] Hadrian, in 120–124 A.D., built the Pantheon we know on the remains of Agrippa's. His rotunda rose on the site of the earlier plaza. The Pantheon's dome is 43 meters in diameter and sits atop drum walls 6 meters thick. The first use of a bronze (metal) roof truss with rivets was in the central entrance portico.

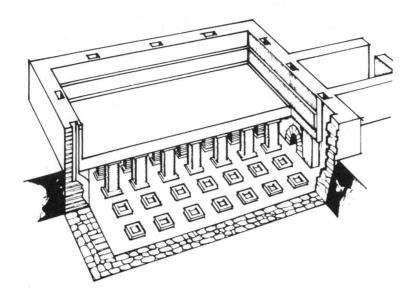

FIGURE 4-11.   Hypocaust and detail of wall conduits. (Redrawn from Sandström, 1970.)

FIGURE 4-12. Interior of the Pantheon (Library of Congress).

NOTES

1. **Anderson, W. J. and Spiers, R. P.** *The Architecture of Ancient Rome.* Books for Libraries Press (1927) p. 17.
2. **Carvill, J.** *Famous Names in Engineering.* Butterworths (1981), p.2.
3. Some writers attribute the use of "Greek Fire" (pyr thalassion) to Archimedes in the defense of Syracuse. This early napalm is more correctly credited to the later Byzantines (ca. 673); see Heath, I. *Byzantine Armies, 886–1118.* Osprey (1979), p. 39.
4. Marcus Aurelius returned to hereditary succession in 161 A.D. setting off the series of "Barracks Emperors".
5. **Rivoira, G. T.** *Roman Architecture.* Oxford (1925).
6. Typical vehicles on Roman roads included: *benna*, cart for several passengers; *carpentum*, two-wheeled gig used by woman; *carrus*, chariot for war or racing; *rheda*, four-wheeled wagon of the imperial Post; *plaustrum*, country wagon with solid wheels; *angaria*, four-wheeled post vehicle; *arcera*, open cart; *birota*, two-wheeled cart(?); *cisium*, two-wheeled gig; *clabula*, ox cart, military transport wagon; *tensa*, show chariot. From Chevallier, R. *Roman Roads.* B. T. Batsford (1976), pp. 178–179.
7. **Delebrück, R.** *Die Drei Tempel in Forum Holitorium.* Rome (1903).
8. **Chevallier, R.** *Roman Roads.* B. T. Batsford (1976).
9. Ibid.
10. **Forbes, R. J.** *Man the Maker.* Henry Schuman (1950), p. 76.
11. **DeCamp, L. S.** *The Ancient Engineers.* Ballantine (1963), p. 194.
12. **Chevallier,** op. cit., p. 28. Sandström relates that Peutinger, a courthouse librarian and historian, was given the map in 1507 for safekeeping, with the admonition to publish it. He died without doing so and the map was not found (again) until 1714, when an inventory was made of Peutinger's library. The map eventually wound up in the hands of Emperor Charles IV (17 years later) and was placed in the Imperial Library. It now resides in the Vienna State Library (cf. Sandström, G. E. *Man the Builder.* McGraw-Hill (1970), p. 151).
13. **Sandström,** op. cit.
14. **Baradez, J.** Réseau routier de la zone arrière du limes de Numide. *Limes-Studien,* II, 59, Bâle (1957), pp. 19–30.
15. **Dilke, O. A. W.** *The Roman Land Surveyors.* David and Charles (1971), p. 66; see also Thulin, C. Die Handschriften deu Coprus Agrimensorum Romanorum. *Abhand Königlich Pressische Akademie der Wiss. Philos. Hist.* Classe, Part II (1911).
16. Ibid.
17. **Hill, D.** *A History of Engineering in Classical and Medieval Times.* Open Court (1984), p. 119.
18. **Dilke, O. A. W.** Mathematics for the surveyor and architect. *Archaeol. Today,* **8**(8): (1987), p. 31.
19. **Columella,** *Agriculture.*
20. **DiFenzio.** Sulla portala degli antichi acquedotti romani e determinazione della quinaria. *Giornale del Geniocivile.* Rome (1916).
21. **Frontinus, Sextus Julius.** *"De aquae ductu,"* The Two Books on the Water Supply of the City of Rome. (97).
22. **Ashby, T.** *The Aqueducts of Ancient Rome.* McGrath (1973), p. 125.
23. Ibid., pp. 36–37.
24. **Biswas, A. K.** *History of Hydrology.* North Holland (1972), pp. 79–103.
25. $1\frac{1}{4}$ digits (.73") = 1.85 cm. 12 digits (8.75") = 22.2 cm.
26. **Van Den Broucke, S.** Aqueducts, cisterns, and siphons. *Archaeol. Today* **8**(4): 18 (1987).
27. Ashby, op. cit., p. 79.
28. **Livy, Titus Livius,** *History of Rome.* I. M. Dent & Sons (1912–1924).
29. **DeCamp,** op. cit., p. 201.
30. **Kirby et al.** *A History of Engineering in Classical and Medieval Times.* Open Court (1988), pp. 62–63.
31. **Straub, H.** *A History of Civil Engineering.* M.I.T. (1964), p. 13.
32. **Hill,** op. cit., p. 68.
33. **Middleton, J. H.** *The Remains of Ancient Rome.* London (1892) p. 9.
34. **Mark, R.** *Mystery of the Master Builder.* Coronet Film & Video (1990).
35. **Ashby,** op. cit., p. 79.
36. **Vitruvius,** *De architectura librum.*

5 ANCIENT POWER AND METALLURGY

During antiquity (Classical Civilization) the available prime movers were men and harnessed animals. Manpower was always available in antiquity. Only in the later Roman Empire do we hear of acute labor shortages which contributed to the introduction of the water wheel and other machinery. This ancient state used statute or corveé labor and slaves. Cranes, wheels, and sledges were worked by gangs of men. Men manned the galleys, built the roads and aqueducts.

Slavery was a recognized social institution in antiquity for providing cheap labor. There is no basis for sweeping statements that "slavery impeded the use and evolution of machinery and engineering."[1] Slavery changed over antiquity and in fact had many forms. In Egypt, slavery played a minor part in the economy until Hellenistic times. In the Near East it was part of the economic pattern from the beginning, but was a low percentage compared to that of the free population. Industry employing large numbers of slaves was not found. The ancients preferred tenants to slaves.

In Greece, slaves were paid the same as freemen in most artisan shops. In Rome, workshop employment of slaves increased during the second half of the 3rd century B.C. Large numbers were employed on Roman estates (*latifundia*), a diversified agriculture of large farms which replaced smaller independent holdings.

200 B.C. to 100 A.D. was the central period for extension of the use of slave labor over free artisans in Italy. Concomitant was a change in accepted Roman attitudes toward life in general and the role of the Roman state in the world. After 100 A.D. saw the beginnings and growth of manumission of skilled slaves due to the humanistic influence of Stoic philosophy. Why did harnessed animals not take over a larger role in labor? The ancients had insufficient knowledge of animal anatomy which saw ox harness used on mules, horses, and donkeys with disastrous effects, thus robbing these animals of their superior tractive power (see Table 5-1). As a result, the horse could exert only a fraction of its available energy with the ox harness design. Instead of pulling 15 times what a man could, the horse could only pull four times that of a slave. The Roman agronomists recognized the relative amounts of food were also 4:1 (horse to slave). Thus, there was no economic advantage to the horse where large pools of manpower were available.

The camel was used by the Assyrians in desert areas but appears rarely in the Mediterranean. Under modern conditions, a horse can do ten times the work of man, much less in ancient times (Table 5-2). The point of harness attachment used in antiquity was in back of the horse's neck with the collar high on the throat instead of lowered to rest on the shoulder blades (Figure 5-1). The collar harness was not evolved until the 7th century A.D. on the Steppes and was not used until the 10th–12th century in Europe. Improper horseshoes impeded the use of donkeys, mules and horses — true iron shoes were not developed until around the 2nd century B.C. in the Celtic area — with the grip on the hoof not seen until much later (9th century A.D.).

MACHINES

Machinery in antiquity was thus either powered by man or water. Lewis Mumford[2] describes the period of technology of Imperial Rome as a "wood and water" complex. Rome and later cultures until the 18th century used wood for building and water as a prime mover. DeCamp[3] tells us the Greeks and other cultures understood the capstan, windlass and crank, thus using rotary motion for work such as grinding, lifting or pumping.

TABLE 5-1
Muscular Power of Man and Various Animals

	Pressure exerted (kg)	Velocity (m/s)	kg/m/s	Ratio
Horse	54.4	1.09	59.7	1.00
Ox	54.4	0.73	39.8	0.66
Mule	27	1.09	29.9	0.50
Ass	13.5	1.09	15	0.25
Man, pumping	6	0.74	4.6	0.076
Man, turning winch	8.2	0.74	6.2	0.104

TABLE 5-2
Tractive Effort of Man and Animals (Net Load Moved Horizontally)

	Net load drawn (kg)	Velocity (m/s)	Load moved (kg/m/s)	Ratio
Horse				
w/cart	680	1.09	1645	1.00
Carrying load, walking	122	1.09	296	0.18
Man				
Wheelbarrow; (2 wheels)	101	0.5	114	0.7
Wheelbarrow; (1 wheel)	60	0.5	67	0.4
Carrying load	40	0.75	68	0.4

Water mills were applied to corn grinding and drove powerful hammer forge and bellows for metallurgy. This contributed to increase in size and quality of wrought iron. Mining hoists were driven by water wheels in some areas. Cog and gear wheels evolved with water wheel technology.

The prime movers were the Norse mill and Vitruvian Watermill (shown in Figure 5-2). A 5th century B.C. poem first mentions the water wheel. The "Greek" mill or Norse mill is the oldest design. By the 1st century A.D. an "overshot" or Vitruvian horizontal mill was in use near Naples with a heavy nave (dia. 74 cm = 10 Roman digits) which carries 18 spokes. The total wheel height was 25 Roman digits. The undershot Vitruvian design had an output of 3 hp @ 46 rpm which ground 150 kg corn/h compared to hand querns operated by two slaves which produced 7 kg/h. The introduction of the water-mill was tied to water supply. The rivers in the Eastern Mediterranean area carry widely varying quantities of water in different seasons and hence do not provide a constant supply. As we have seen, aqueducts were necessary for constant supply. Mills in western Europe were more plentiful. The rivers were more constant in supply.

An idea by a Latin-speaking inventor in *De Rebus Bellicus* (366–375 A.D.) illustrates a warship or "Liburna" with three sets of oxen turning a capstan rotating two paddle-wheels outside the ship — his invention precedes the Chinese claim of having used human-moved paddle-wheels of the 12th century. Floating mills were invented in 537 A.D. by Belisarius when Goths cut Roman aqueducts.

Hero of Alexandria describes a wind organ with a piston providing air for the organ pipes moved by a wheel. Hero is the only classical engineer who mentions wind power in his writing. Other works by Hero that specifically discuss machines include *Mechanics; Siegecraft; Automation-making*; and *Dioptra*.[4] In these he describes means for achieving mechanical advantage. By combining vertical spars, a windlass and lifting tackle the sheerlegs crane was utilized in raising loads to height. Vitruvius[5] describes one (Figure 5-3) and others are shown with a treadwheel or squirrel-cage-like drums for power. Combined with com-

FIGURE 5-1. Throat girth and collar harness, Roman period (U.S. Bureau of Roads).

pound pulleys with ratios of 2:1 and 3:2 great weights of column drums or blocks could be raised.

The Greek engineers along with the earlier Assyrians and the later Romans were first and foremost masters of the machines of war. The heliopolis or "city destroyer" was first used by the Assyrians to besiege and reduce walled towns of Mesopotamia. It was later used by Phillip, then Alexander of Macedon in campaigns in their conquests of Greece and Persia.

Philon of Byzantium[6] in his *Elements of Mechanics* discusses catapult designs, his and those of the earlier Dioysios and Etesibios. The origin of these machines is variously attributed to either the Grecian Syracuse or the Phoenicians of Tyre and Sidon.[7]

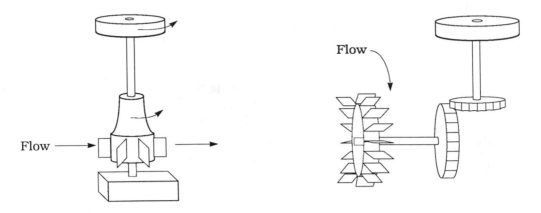

FIGURE 5-2.   Norse and Vitruvian water mills.

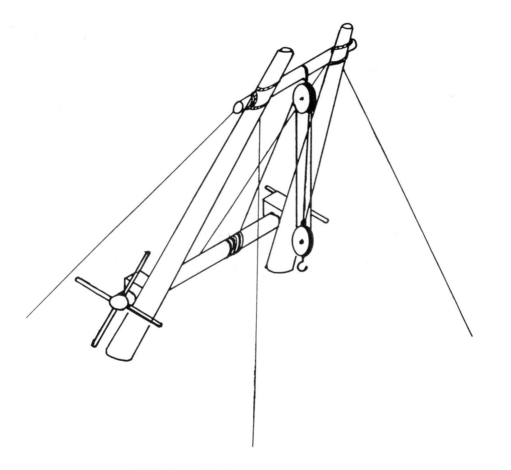

FIGURE 5-3.   Sheerlegs. (Redrawn from Hill, 1984.)

Termed as either dart throwers (Greek, *oxybeloi*) or stone throwers (*lithoboloi*), catapults, as a class of machines, evolved from the 4th century B.C. to remain a mainstay of armies until the invention of gunpowder. The catapult or more properly *ballista* is mentioned at Syracuse's defeat of her arch-rival for Sicily, Carthage, at the siege of Motya, a colony of the latter in the west of that island.[8] Developed from the principle of the flexion bow, the catapult was designed around a wooden beam with a trough or *syrinx* in its upper surface.

A bow was fastened to the forward end of the beam. The bowstring was cocked by sliding a smaller beam (*diostra* or projector) to which it was attached back to the crosshead with a trigger mechanism, typically a hook lever. To pull the bowstring a windlass was used. The projectile or dart-like spear was loaded in the *syrinx*. The range of these *ballistae* is unknown but later torsion designs could fire meter-length arrows over 300 meters.[9] These torsion machines replaced the tension bow with a pair of massive winds of rope of hair ("skeins") mounted on either side of the *syrinx* beam. Into the skeins were placed rigid arms which attached to the bowstring. These *ballistiae* could launch lead or stone balls as well with nearly the range of the arrows.

The torsion principle was extended by Roman artillery engineers to the design of the *onager* or "mule". This weapon was a lever-arm device with a sling attached for projectiles such as stones. The arm was relatively vertical inserted into a skein mounted horizontally in a frame. Pulling back the arm with a windlass cocked the weapon. Reconstructions firing stones weighing 1 to 4 kilograms could reach a range of 300 meters. Such a distance requires tensions of over 7000 kg cm^{-2} in the torsion winding[10].

Wind Power in Antiquity — The Ship as Machine

Of the machines of antiquity none conforms better to Mumford's definition of "wood and water" technology than the ancient ship. It was the only ancient machine to utilize the wind as a motive force, albeit poorly. Ancient ships were notoriously poor sailers. Square-rigged, steered by oars, round-bottomed with shallow draft, these vessels could only sail effectively with the wind dead astern. With their primitive sail designs and rigging their operators could not follow any shift of wind that moved far up their beam. Even if they had been able to do so their round, shallow hulls would not have let them take advantage of it as they would slide sideways. To avoid these disadvantages the builders turned to their old standby — muscle-power.

Oared ships were favored as warships by the Egyptians by Ramses III's time (1198–1166 B.C.). The Phoenicians became the premier seafarers of the Mediterranean with their ships departing in design from the earlier Egyptian models. From Cyprus they probably acquired the "keel and top-sails". Banks of oars were added along the vessel's side and became the most reliable form of propulsion. Only on warships could the cost of rowers be expended. Even if they had been slaves, as popularly believed, no master of a commercial vessel could afford to feed them, as we recall from our example with the economics of man vs. horse. Merchant ships evolved along a different course in antiquity with the warships specializing in the multi-banks of oars, high length-to-width ratios ($\geq$7:1), and prows designed for ramming. Even then, both classes of vessels were capable of being beached as much of the Mediterranean lacked true harbors. Ships were drawn up on shore to load or to repair. This feature allowed the Greeks to design the portage road at the Isthmus of Corinth without extensive dock or handling facilities. The vessels "beached" onto cradles for the tow across the Isthmus.

In the warships more rowers were added. Hull length and weight largely determined how large these galleys could realistically become, so the design resorted to rowing arrangements of multiple banks of oars and rowers. The first arrangement was bi-level with three men on the oars. It was termed a triere or trireme (Figure 5-4). Modern reconstruction of this classical design has reached the 7 knot speeds claimed for it.[12] In the heyday of the Athenian navy the trireme was built to dimensions around 40 meters long by 7 meters wide.[13] As many as 200 men served aboard the trireme, exclusive of marines, with 170 doing the rowing. These vessels were the principle naval warship until Hellenistic times.

Archimedes designed a huge galley named the SYRAKOSIA. This vessel was the first three-masted ship in the world. It was an anomaly which was too big for most harbors, too slow for battle, and too uneconomical for a merchant vessel. As an engineering landmark,

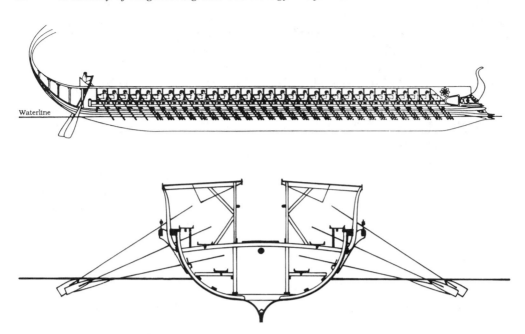

FIGURE 5-4.   *Upper:* Elevation of 5th Century B.C. trireme. (Courtesy of L. Casson) *Lower:* Oar arrangements. (Courtesy of J. F. Coates)

it is perhaps interesting because of the genius of its builder rather than its technical features. It added nothing to nautical design that was not already evident in the classic trireme. Still the classical ship that outstripped even SYRAKOSIA for size and overdesign was the war galley of Ptolemaios IV Philopater, ruler of Egypt from 221 to 205 B.C. This vessel (Figure 5-5) was 128 meters long and 42 meters wide with double bow and stern mounting seven rams.[14] Four thousand men rowed her pulling 17 meter-long oars. Perhaps its legacy is being one of the first examples of the answer to a question that should not have been asked.

Rome, practical as always, relied on the standard warship design in the quinquireme or pentere with a single row of five man oars per side to sweep Carthage from the sea. In their battle against the Celtic Veneti of the Brittainy coast they met vessels that truly sailed in battle. According to Julius Caesar[15] these ships had leather main sails, probably square-rigged, that allowed them to out-run the Romans. Their defeat came only when the wind failed them at the crucial engagement with the Roman fleet.[16] As these designs "lost" in the Roman view of things they were never brought into the Mediterranean. We cannot know if later Scandinavian designs owe anything to the Celtic ships, although we do recognize a distinct design tradition in the northern European world, even during Rome's dominance.

The keel is the "back" of the wooden boat — functioning much like the skeletal counterpart in articulating the attached "vertebrae" (frames). This design originated in the Northern Bronze Age in Europe which began about 2000 B.C. On the Hjortspring Boat, a Bronze Age vessel from Denmark, the bottom plank extends forward of the hull to form a "cutwater" or break. The stern and stern posts were joined to this plank. A few wide, thin strakes form the hull which had thin ribs braced with struts for transverse solidity at gunwale level. The Hjortspring craft was 13 meters long and 1.8 meters in breadth. She was paddled and carried no sail.

The ship design of the North European waters was a radical departure from that seen in the Mediterranean. The Nydam boat is an early step in this separate tradition — the "clinker-built ship". Clinker-built refers to the practice of over-lapping the hull strakes or planking rather than edge-joining the strakes (carvel-built). By 600 A.D., vessels built in

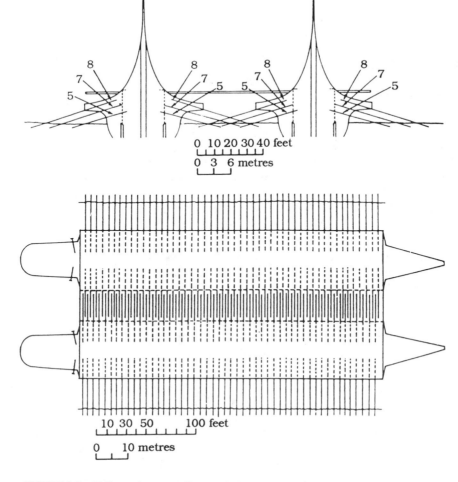

FIGURE 5-5. Philopater's super galley, ca. 3rd Century B.C. (Redrawn from Casson, 1986.)

Norway had true keels rather than center planks. Further, they were designed and built to carry sail by 800 A.D.

EARLY METALLURGY — COPPER

The development of metallurgy was first assumed to be the province of the cultures of Mesopotamia. Recent studies have revealed two related facets of early metallurgy: (1) it was obviously originated in different places in the Old World, and (2) the higher cultures of antiquity (Sumer, Egypt) did not invent metal working.

Evidence in Europe points to metal-using sites as early as 4500 B.C. along the Danube. Other early sites occur at Los Millores (Spain), northern Italy, Ireland (Mount Gabriel) and Anatolia (Halicar and Çatal Hüyük).

The Vinca culture flourished in what is now Yugoslavia around 5000 to 3000 B.C. This group probably utilized copper ore deposits in Transylvania, Bosnia, Serbia, and Macedonia. At Rudna Glava (Bulgaria), archaeologist Boreslav Jovanovic has found vertical mine shafts up to 30–40 meters deep. The miners obviously followed the ore seams using antler picks and fire/water mining methods. Early copper technology and hence metallurgy was developed by various technical specialists in these early Balkan and European cultures. The task was not easy as copper had a high melting point — 1080°C. Experiments to try and reproduce pure copper metal using ores and materials available to early metallurgists have met with mixed results. Perhaps the use of compounds or alloys — copper-antimony, copper-arsenic or copper-tin — creating a eutectic (lowered melting point) was discovered early by these specialists. With present evidence this is purely conjecture.

Metallurgy — Copper to Bronze

The production of copper by the ancients required a pyrotechnology that could smelt the copper-bearing ores such as malachite $CuCO_3$ $Cu(OH)_2$, chalcocite Cu_2S and chalcopyrite $CuFeS_2$. As these are low-grade (less than 2% copper) the copper must be concentrated by the smelting procedures. In some instances "native" copper or high grade metallic ores such as those found in the Great Lakes area were "cold-worked" into crude tools and weapons.[17] This occurred in both the New World and the Old World. The antiquity for the Lake Superior mining dates to the Middle Archaic Period (5000 – 3000 B.C.) and that for Anatolian region experimentation to as much as the ninth millennium B.C. Sites such as Çatal Hüyük in Anatolia have native copper implements in the 7th century B.C.

By the early 4th millennium, copper smelting was used to produce tools in Anatolia such as knives, sickles and woodworking adzes.[18] Copper metal in pure form is quite malleable and has decidedly less tensile strength (≤ 3000 kg cm^{-2}) than stone or the latter alloy of copper-bronze. This tensile strength was enhanced by hammer working of the tool after casting. To produce the copper the smiths roasted the sulfide ores to oxidize the sulfides of impurities such as arsenic (As) and antimony (Sb). Cu_2S remains as a calcine. The carbonate ores presented less of a problem to the smelters and were processed in the simple bowl-shaped hearths to form "puddles" of 98–99% pure metal. Because of the early primitive smelters and generally large amounts of ore needed, the metal produced was utilized for exclusive items such as jewelry or small tools.

As smelting techniques were extended in intensity and efficiency the demand for ore increased. It was probably inadvertent experimentation with various ores that led to the discovery of bronze in either of two forms — arsenical or tin bronze.

Arsenical copper ore such as enargite, Cu_3AsS_4, produces a natural bronze alloy.[19] This metal improves the tensile-strength properties of the alloy to roughly double that of the unworked pure metal. It is possible to form tin-bronze from stannite, Cu_2FeSn_4, by direct smelting.[20] As either an accidental alloy or, by 3000 B.C., an intentional metal process, bronze was the predominant metal until the development of iron as a useful metal by 1000 B.C. The transition from arsenical copper to tin bronze is demonstrated in the following figures showing percentages of copper-based artifacts containing more than 5% tin for Europe and the Near East for the periods, ca. 2700–2200 B.C. and 2200–1800 B.C. (Figure 5-6).

Major tin sources in Europe and the Near East include Cornwall (Britain); Spain; and Turkey (Anatolian Plateau). Sources for Egyptian tin have been identified in the Sudan during the reign of Pepi II (22nd century B.C.). Outside Europe and the Near East, major Bronze Age cultures are known to have existed in Thailand (Bai Chang) and Northern China (Hunan — The Shang Dynasty ca. 1400 B.C.).

The development of metallurgy comes in the Neolithic Period and appears in Europe (Spain, Balkans) and the Near East (Anatolia) as well as in Southeast Asia. While contemporaneous with ceramic manufacture as a technology in some areas, such as Anatolia, the early metallurgy seems to have actually preceded the production of pottery. Bronze was utilized in building as fasteners and even for truss in the Pantheon. Beyond these uses the metal was used primarily in weaponry and tool manufacture.

Iron Metallurgy

The first smelting of iron may have taken place as early as 5000 B.C. in Mesopotamia.[21] Throughout the Bronze Age, iron was produced sporadically. Iron droplets, a by-product of copper smelting, formed lumps on top of the slag from the smelting process. This was

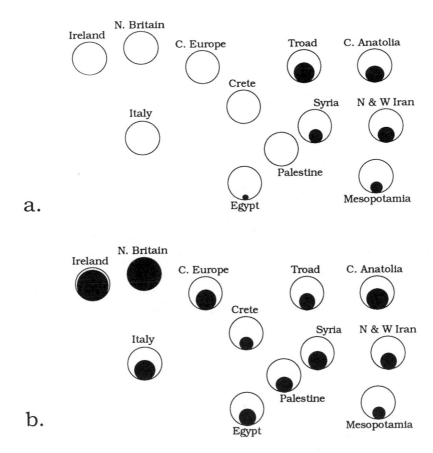

FIGURE 5-6.   (a) Percentages of bronze artifacts in Europe and Near East, ca. 2700–2200 B.C.; (b) percentages of bronze artifacts in Europe and Near East, ca. 2200–1800 B.C. (From *The Search for Ancient Tin*, Franklin, Olin, and Wertime, Eds., U.S. Gov't. Printing Office, 1977.)

due to the increased efficiency in copper smelting by the use of iron oxides with the copper ores, particularly the carbonate ores. The two reactions in the process are

$$(CuCO_3)CuOH_2 \xrightarrow{\text{heat}} 2CuO + CO_2 \uparrow + H_2O \uparrow$$

then the CuO is reduced by carbon monoxide,

$$CuO + CO \xrightarrow{\text{heat}} Cu + CO_2$$

Silicate impurities in the ore react with the CuO to form the slag

$$CuO + SiO_2 \longrightarrow (CuO)_2 \cdot SiO_2$$

The iron oxides in the copper ores increased the copper yield thusly

$$(CuO)_2 * SiO_2 + Fe_2O_3 + 3CO \longrightarrow 2Cu + (FeO)_2 + 3CO_2$$

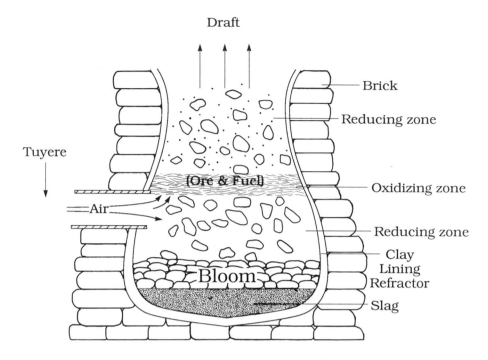

FIGURE 5-7.   Bloomery Furnace.

To reduce iron in large amounts required a CO/CO_2 ratio of 3:1, whereas only 1:5000 CO/ CO_2 was required for the reduction of CuO. Forging these lumps of early iron was difficult as the iron contained copper and sulphur. By 1200 B.C., the smiths had begun to roast Fe_3O_4 in smelting furnaces. In this process, at 1200°C, the iron did not melt as did copper, but only became pasty, as iron requires 1540°C to melt.

The production of iron is not a melting operation, it is a reducing operation. The object is to reduce Fe_3O_4 to metallic FeO. Today, we accomplish this with blast furnaces but in the early days the smiths roasted the iron ore in a forge called a ''pit forge'' or later a ''bloomery forge'' (Figure 5-7). To achieve the required temperature, some form of device was needed to create a draft. The first smiths utilized blow pipes, as illustrated here. The result of this crude smelting was a pasty mass of iron, slag and charcoal called a ''bloom''. With further heating and hammering at 1250°C the bloom could be consolidated into a wrought iron implement. When hammered, the fluid slag and oxides were squirted out and the iron particles welded together. This slag ''gangue'' was termed fayalite.

Bloomery iron is a poor substitute for bronze — its tensile strength is about 40,000 psi compared to the strength of pure copper at 2250 kg cm^{-2}. Hammering brings the strength of iron to almost 7000 kg cm^{-2}. A bronze with 11% tin, however, has a tensile strength after casting of 4500 kg cm^{-2} and a strength after cold-working of 8500 kg cm^{-2}. Further, bronze could be melted and cast at temperatures reached by early furnaces. Iron could be cast only after adding as much as 4% carbon to the bloom, which in turn caused the cast metal to be extremely brittle. Bronze corroded little, iron rusted easily. Why did iron, then, replace bronze? One theory suggests that iron properly carburized (the addition of carbon) produces an alloy with desirable cutting and durability properties. This alloy is steel.

A carbon content of 0.2 to 0.3% gives steeled iron a strength equal to that of unworked bronze; raised to 1.2%, the steeled iron has a tensile strength of 140,000 psi. If the blacksmith then cold hammers the steeled iron, a tensile strength of up to 245,000 psi can be obtained — more than double that of cold-worked bronze! In the bloomery furnace the fuel and ore charge passes down the stack where at 800°C CO reduces the Fe_3O_4 to FeO flakes. With rising temperature the metal agglomerates and forms wustite, which reacts with SiO_2 to form a wustite-fayalite slag. This seals off the iron particles from furnace gases and allows the particles to pass the high temperature, oxidizing zone above the tuyeres. With high grade ores, SiO_2 must be added to form the slag. Below the tuyeres, the furnace atmosphere is reducing again. Slag flows to the bottom of the furnace, with the bloom forming a layer above the bowl. The bloom is ready for the forge where it can be steeled by the introduction of carbon. The carbon was picked up by the iron from the white hot charcoal at 1200°C.

The diffusion of carbon into the iron is temperature dependent — after 9 h at 1150°C, the concentration of carbon is 2% at 1.5 mm. In modern metallurgical terms, the carburized iron has a microstructure known as austenite. As the temperature falls to 727°C, the austenite breaks down into ferrite, pure iron and iron carbide or cementite. The reaction is called the Eutectoid Reaction. The new microstructure is called pearlite, with alternating layers of ferrite and cementite. If the iron contains 0.8% carbon the entire (100%) microstructure will be pearlite.

The ancient smiths developed a method to harden the steeled iron even further. This method was quenching or rapid cooling in air or water. Rapid cooling produces the microstructure called cartensite which is hard but brittle. A third technique for manipulating the end result of forging steeled iron was tempering. When the ancient smiths realized that quenching made hard but brittle steel, they learned the process of tempering, where the steel was heated to the temperature of transformation, 727°C. The iron carbide (in the cementite) precipitates and coalesces, increasing ductility of the metal by creating a finer-grained pearlite on the cutting edges of tools and weapons.

Beyond weapons and fasteners neither copper alloy nor iron metallurgy was important to ancient engineering. Sostratos used iron in the armature for the Colossus of Rhodes; Hadrian used bronze in the truss for the Pantheon's portico; other builders used them as tierods and clamps but structures of metal were not a component of their engineering. The hydraulic engineers recognized the virtues of cast bronze or brass pipes but preferred cheaper lead.

Metal was plentiful in antiquity and relatively easy to mine in surface or near-surface deposits of metallic or oxide copper, gold, and spathic iron.[22] Celtic tribes of Trans-Alpine Europe became master smiths producing tons of iron in the form of mails, spikes, weapons, chariot tires, and other implements. To the east in what is now northern China, the Shang culture (1850–1100 B.C.) was smelting bronze on a par with Anatolia and by the 6th century B.C. the Chou Dynasty was casting iron. Still, like their Mediterranean counterparts, metal was a staple for others than engineers. Mumford's ''wood and water'' terminology may be the correct styling of power and metallurgy in Antiquity.

NOTES

1. **Forbes, R. J.** *Man the Maker*. Henry Schuman (1950).
2. **Mumford, L.** *Technics and Civilization*. Harcourt, Brace and World (1963).
3. **DeCamp, L. S.** *The Ancient Engineers*. Ballantine (1963).
4. **Hero.** *Di Herone Alessandrine. De gli automati*. G. Ponno (1589); *Heronis Mechanici Liber de Machinis Bellicis*. F. Franiscium (1572); *The Pneumatics of Hero of Alexandria*. B. Woodcraft, Ed. Taylor Walton and Maberly (1951).
5. **Vitruvius.** *De Architectura* (2 vols.). Loeb Classics, London (1970).
6. **Philon.** *Les Livres des Appareils Pneumatiques et des Machines Hydrauliques Par Philon de Byzance*, in Notices et Extraits des Manuscrits de la Bibliothèque Nationale, Tome 38 (1903).

7. **DeCamp, L. S.** *The Ancient Engineers*. Ballantine (1963), p. 106.

8. **Diodorus, Siculus.** *The Library of History*. Davis (1814).

9. **Forbes, R. J.** *Man the Maker*. Henry Schuman (1950), p. 83.

10. Ibid.

11. **"Phonecia",** *The American International Encyclopedia*. The John C. Winston Company (1954).

12. **Martin, L.** Ancient Greek warship sails again. *Benthosaurus*. No. 25 (1989), p. 2.

13. **Casson, L.** *Illustrated History of Ships and Boats*. Doubleday (1964), p. 35.

14. **DeCamp,** op. cit.

15. **Brady, S. G.** *Caesar's Gallic Campaigns*. The Military Service Publishing Co. (1947), p. 57.

16. **Brady,** ibid., p. 59.

17. **Fagan, B. M.** *People of the Earth*. Little, Brown (1980), p. 128; Nastian, T. J. Prehistoric Copper Mining in Isle Royale National Park, Michigan, Museum of Anthropology, Ann Arbor: University of Michigan (1969).

18. **Kunç, S.** Analyses of Ikiztepe Metal artifacts. *Anatolian Studies*. XXXVI, (1986), p. 99.

19. **Charles, J. A.** The development of the usage of tin and tin bronze: some problems, in *The Search for Ancient Tin*. Washington, Government Printing Office (1978), p. 28.

20. **Wertime, T. A.** The search for ancient tin, the geographic and historic boundaries, in *The Search for Ancient Tin*. Washington, Government Printing Office, (1978), p. 28.

21. **Vander Merwe, N. J. and Avery, D. H.** Pathways to steel. *Am. Sci.* **70** (2): 146–155 (1982).

22. **Sandström, G. E.** *Man the Builder*. McGraw-Hill (1970), p. 267.

6 BYZANTINE AND ISLAMIC ENGINEERING — PICKING UP A FALLEN TORCH

After 476 A.D. and the fall of the Western Roman Empire, Europe, the Middle East, and Africa became separate power blocks. These were:[1]

- The Kingdom of the Franks (481–814 A.D.) — France-Germany
- The Kingdom of the Visigoths (507–711 A.D.) — Spain
- The Kingdom of the Ostrogoths (493–526 A.D.) — Italy
- The Kingdom of the Vandals (429–533 A.D.) — Africa
- The Eastern Roman or Byzantine Empire

Constantinople was the civilized center of the world, the Byzantine capital. Under Justinian (527–565 A.D.) and his great general, Belisarius, most of the west was retaken by the Byzantines by 565 A.D. They destroyed the Ostrogothic Kingdom of Theodoric (493–526 A.D.) and that of the Vandals in Africa (522 A.D.).[2]

The Visigoths in Spain gave ground but survived. This "Roman Reconquest" was not to last, as we shall see. While the light from this burst of Roman energy lasted, the wisdom of Constantine's founding of Constantinople was fully justified. Indeed, for several centuries this Greek-speaking metropolis would survive, holding in trust, as it were, much of the accumulated wisdom of antiquity. For a brief time the engineers, architects, and scientists who were heirs to this tradition added to it, but as the centuries mounted, scientific and technologic innovation waned as the torch lit anew by Justinian guttered and went out.

The final quenching of the Roman civilization was not sudden. The Byzantine Emperors survived for nearly a millennium after the fall of the West. They did this by virtue of combining diplomacy, out-right deceit, and military preparedness. At the heart of her military strength lay the work of her engineers.

Justinian's great offensive thrust was the last attempt by the Byzantine Empire at real imperialism. By and large she became a defensive power who could withstand siege and survive her enemies, principally the Arabs and later the Turks.[3] Under Justinian over 700 fortifications were constructed or rebuilt by his engineers.[4]

The walls of Constantinople itself rose to over 12 meters and were 4 meters thick with a core of concrete.[5] The construction was ashlar over concrete, faced in places by brick, mostly as decoration. Walls at Dara on the upper Tigris River were 18 meters high and almost 10 meters thick at their base.[6] At Nicea the double enceinte or two-wall design was used, with the outer wall 2 meters thick and the inner wall 4.5 meters thick. Cairo was fortified and Rome's nine walls were strengthened and modernized. Still, it was Constantinople that would have to ultimately stand the pressure of Arab sieges in the 7th and 8th centuries (668–675 A.D., 716 A.D.).

The great walls could keep the invader out but siege was not designed to overtly conquer by force of arms. The aim was to allow starvation and thirst to do that for the besieger. Constantine's builders had seen to the hydraulic needs of the city by building aqueducts like those of Rome. Cisterns below the city, begun by Justinian, could hold over 34 million liters.

Great wealth and power concentrated in Constantinople.[7] As political and spiritual center of the late Roman period it was natural for Justinian to match his conquests with monuments.

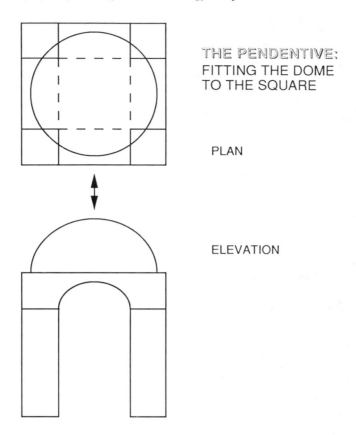

THE PENDENTIVE:
FITTING THE DOME
TO THE SQUARE

PLAN

ELEVATION

FIGURE 6-1.   The Pendentive was a device used by Byzantine architects to erect a round dome on the square central area of a church. In effect they put a hemisphere on a square box and cut away the hemisphere's crown and overlapping sides on which a smaller dome or hemisphere could rise.

His most enduring construction is that of Haghia Sophia, the great domed church built between 532 and 537. Like the Pantheon it used the dome but unlike that older structure, Haghia Sophia's dome was of lighter build and supported differently. The builders of this church were Anthemius of Tralles and Isodorus of Miletus whom Procopius refers to as "mechanike" and "mechanaporos".[8] The translation of these terms is "engineer".[9] These men used a plan that reflected the state, centralized and awe inspiring. The dome rested on a square. This differs from the Pantheon which is on a drum. To rest a dome on such a plan called for the invention of what is termed the pendentive dome (Figure 6-1). Seen also in partial cross-section (Figure 6-2) this dome is that of Isidoros the Younger, who rebuilt the church after an earthquake damaged it in 558 A.D.[10] The great height of Haghia Sophia comes from the use of 30 meter columns topped by 18 meter arches which support the dome. Erected over an area of 1300 square meters, one can appreciate the impression of such a space. It was built using tried and true Roman materials — brick and concrete.

Byzantine control of the Mediterranean world was to falter. Much of Italy except for the South and Ravenna, was lost 2 years after Justinian died (567 A.D.). By 620 A.D. most of Spain was retaken by the Visigoths and by 697 A.D. the North African provinces were lost to a newer more permanent polity — the Arabs. It was against these peoples and their later successors the Turks that the Byzantines were to conduct their long wars. In these conflicts the use of military engineering was the key to Byzantium's survival. The use of chemical weapons, particularly "greek fire", proved pivotal in defense of fortifications and at sea.[11] Eventually the Arabs learned the secret and used it on invading Christian armies of the 11th and 12th centuries.[12]

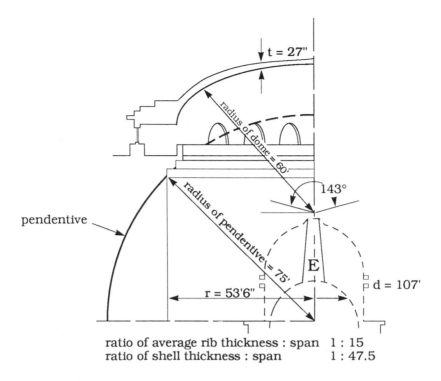

FIGURE 6-2.   Haghia Sophia Dome, Constantinople. (Drawing courtesy of Professor Daniel F. MacGilvrey.)

ISLAM — KEEPER AND BUILDER OF TRADITION

Muhammed (570–632 A.D.), a trader turned prophet, saw visions of the angel Gabriel. He taught a policy of tolerance, with no discrediting of other religions. Fanatical Arab armies conquered Syria and Mesopotamia by 638, Egypt by 640, Persia by 641, North Africa in 697 A.D. and by 711 conquered Spain, invading France. They besieged Constantinople, but were beaten off, in part by Byzantine's "Greek Fire".

The Golden Age of Islam was from 900–1100 A.D. During this period the Arabs founded centers of learning, copied Greek texts into Arabic, developed experimentation to test theory, developed books from paper (University of Cordoba had 600,000 titles in its library in 900), developed the horse collar, windmill, water wheel, and astrolabe, and developed "Arabic" numerals including zero and decimal point.

These mounted, nomadic armies swept out of the Arabian deserts with no baggage trains to encumber them. There were no infantry to slow them so their seemingly lightning conquests become more understandable. They bowled over or bypassed Byzantine fortifications throughout Northern Africa and the Middle East. Combined religious fervor on the Arab part and war weariness on the Mediterranean world's part also contributed to the rapid advance that continued after the Prophet's death (632 A.D.).

The Islamic world was a melting pot of numerous cultures. After the conquest period (ca. 750), the civilization that was to advance the learning of antiquity began to form. To add to this mix came expatriate Byzantine scholars fleeing religious disputes within that empire. The first Caliphs, of Damascus and later of Baghdad, sent to Alexandria and Byzantium for manuscripts to build their libraries. With the conquest of Samarkand (752 A.D.) the art of paper making came into their hands. By 793 the first paper mill was operating in Baghdad.[13]

Arab scientists and engineers used the books on which the translations were written and went on to improvements, refinements, and inventions based on this knowledge. Empirical experimentation was not frowned upon as it was in Hellenistic tradition. Laboratories and workshops were sources of inquiry. Combined with mathematical and algebraic thinking from India and the engineering of the West the Islamic practioners such as Muhammad-Ibn-Musa went on to develop concepts for the zero and the decimal system.[14,15] They developed the astrolabe for measuring angles and celestial positions. Writers like al-Kindí (897–961) wrote over 200 works on meteorology, optics, and force in the 9th century; al-Hazen (965–1039) wrote on optics and perspective; al-Razi, over 100 medical works; the Banū Mūsà (literally "Musa brothers") and al-Jazari, who wrote on automata and water-raising machines, appear in numerous books and manuscripts.

Arab commentators enlarged little on Euclid's geometry but they developed spherical trigonometry, and invented sine, tangent, and cotangent. They developed the pendulum and in physics studied optics and astronomy. In medicine they used anesthetics and surgery. Modern chemistry began with the Arabs—"alcohol" is an Arabic word. Fertilizers were used, as they farmed scientifically. Their architecture was Syrian, Egyptian or Persian. The horseshoe arch developed on the pointed arch of Syria and Egypt as a more efficient structural design. Geometric patterns prevailed over Hellenic realism in their art and decoration.

Universities were founded in Basra (Iraq), Kufa (Iraq), Cairo, Fez (Morocco), Cordoba, and Toledo. Cordoba by 900 had over 600,000 books in one library. Toledo, former seat of the Visigothic kingdom,[16] boasted over 400,000 titles. Arab science invented a chemistry that we can recognize with evaporation, filtration, sublimation, melting, distillation, and crystallization. They developed sugar refining, alcohol, sulfuric and nitric acids, as well as the production of gasoline.[17,18] The Arabs (ca. 640) built windmills to pump water from wells for irrigation. They also ground corn with windmills. The mills had iron shafts which were utilized for rotary axles with vanes on the lower wheel and grinding wheels above the vane. They were typically oriented to the north wind, with shutters to control the venturii.[19]

The origin of these windmills was Persia and Afghanistan. There the wind is constant, blowing harder in the summer months. The builders attached shutters to the openings to control the airflow and thus regulate the rotary speed of the mill. The Arabs also built cooling towers in which they placed wetted sails of cloth. These towers, attached to larger buildings such as mosques, provided humidity and cool air to the interiors. The windmill moved east to China and west to Europe, being almost nonexistent at its place of origin due to the destruction wreaked by the Turks and Mongols in the later 14th and 15th centuries. The Arab engineers were particularly drawn to civil works that involved water. Originating in countries where water is at a premium, these engineers became masters of hydraulics. They refitted ancient systems and built new ones. To power these irrigation and water supplies, they typically used water wheels. They also built diversion dams and used the ancient qanat technology of Persia.

Noriae, or undershot water wheels, were used to hoist water in Hellenistic times.[20] Water wheels are discussed by Vitruvius as well.[21] Arabic authors translated these works, making additions as well.[22] Commentators such as Ibn al-'Awwān, who wrote in the 12th century, understood the principle of the flywheel.[23] It is perhaps this capacity for storing and translating energy that drew the Arabs to the Noria.

Water has a deceptive force and it is this, coupled with its steady, almost timeless constancy, that makes it so appropriate for mechanical tasks. At first, it was used for irrigation and water supply such as that of the great Syrian site of Hama on the Orantes River. The largest noria at Hama has a diameter of 20 meters and produces 80 l/s or a quarter million liters an hour.[24] Shown in Figure 6-3, the noria was an undershot design of relatively low efficiency (∼22–33%). The noria depended exclusively on the flow of current for its motion and lost much momentum due to drag and friction of the immersed component of the wheel. Add to this the inertia inherent to the large wheel and one can see the reasons for low efficiency. Russell calls these losses in efficiency hydraulic and mechanical.[25] This effi-

Noria

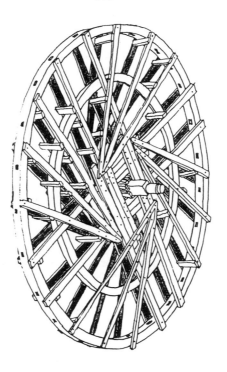

Detail of Noria

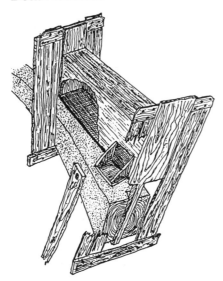
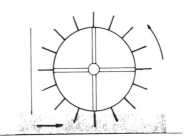

FIGURE 6-3.   Noria. (After Hill, 1984.)

ciency, in modern terms, is the ratio of the power derived at the shaft to that of water at the wheel. Efficiency is independent of head, depending only on bucket or paddle angle, wheel speed, and incident flow (velocity). Nonetheless, it was steady and needed little or no maintenance over a 24-h period.

Having read the authors like Philo, Hero, and Vitruvius, the Arab engineers appreciated all varieties of the wheel besides the noria. They clearly utilized the overshot wheel, the horizontal wheel, and probably had a variant of the overshot—the breast wheel—although this is not evident in their writings. They used the ship and tidal mills. To increase the rate of flow to the wheel the engineer would set the mill between bridge piers to take advantage of the higher current velocity through the opening.[26]

Dams to create reservoirs for water supplies were an Islamic innovation.[27] The Romans and Greeks had built dams which the later Arabs repaired and used. In Syria, again, a large Roman dam still is used on the Orantes near Homs. It was built in 284 by the Emperor Diocletian, forming a reservoir 15 kilometers long and over 5 wide. It was of rubble construction bound by hydraulic mortar faced with basalt. The dam is 2 kilometers in length and about 7 meters wide and the base 20.[28] It came under Muslim use by the 8th century.

These engineers built dams on rivers in all the areas of their realm from Iran to Spain. They built three dams across the Tigris alone and in Iran 'Adud al-Dawla built a dam called the Band-i-Amir in 960 on the Kur River. The dam was of masonry set in concrete and reinforced with lead dowels. These dowels were lead poured into cut grooves to fasten the adjacent blocks.[29] The dam provided irrigation supplies for 300 villages. It stands today, being about 10 meters high and over 70 long.

In Spain, the western Muslims, called "Moors" by the Christians, were building extensive hydraulic systems in what had been the Kingdom of the Vandals. This is translated into Arabic as "al-Andalus" or land of the Vandals, modern Andalusia.[30] Since the area had been under Visigothic control since the 6th century one has to wonder at the Moor's choice of names. Spain or al-Andalus did not bloom figuratively or literally until the 10th century with the coming of the Umayyad Caliphate. The Umayyads had themselves lost control of the eastern Islamic realm to the Abbasids in 749, who moved the capital to Baghdad. Fleeing the fall was one 'Abd al-Rahman, a surviving prince of the Umayyad dynasty who came to Spain in 755.[31]

Cordoba became the capitol of Moorish Spain. The Umayyads brought Syrian engineers to transform al-Andalus from a backwater to the brightest jewel of the Islamic world. Transplanting hydraulic techniques developed at Homs and Hama, these men changed the Iberian peninsula. Irrigation brought new crops such as rice and sugar cane.[32] Other crops such as melons, fruits, cotton and nuts flourished.[33,34] Bathhouses, fountains and water gardens were built in all the Moorish cities to include Toledo, Seville, Granada, Valencia, and Barcelona. In Cordoba alone were built 300 public baths.[35] The Moors utilized existing Roman systems such as the Segovia, Toledo and Almunecar aqueducts. Unlike the Romans, Islamic engineers did not separate the supplies for irrigation and water supply. Both used the same sources. Dams impounded water for both purposes. One notable example which still exists at Murcia is 128 meters long and 7.6 meters high. For three quarters of its length it is 48 meters thick at the base and 38 for the remainder.[36] This means of construction was deemed necessary by the builders to prevent foundation problems since the river bed was soft. The engineers used the weight of the dam to resist uplift and water pressure. It is a rubble-core dam faced with masonry. Smith[37] notes the builders' appreciation for the force of falling water by their building a platform 4 to 5 meters below the crest to dissipate the energy of water coming over the dam. The overflow then ran to the foot over a flat or gently sloping face, causing the dam to act as a spillway. The risk of erosional scour caused by a head of nearly 8 meters was thus reduced.

Like the Byzantines, the Arabic builders lavished their greatest efforts on their churches or "mosques". The world of the early Middle Ages was one of great religiosity. Christianity's bastion was the Byzantine Empire and its great rival was Islam. As in all things practiced with great fervor, the friction along their common border created flashpoints that led to constant conflict. As we have observed, religious disputes within Christendom generally benefited Islam as scholars, scientists and engineers fled to the more liberal fiefs of the Caliphate. Still, wars came due to religious and political energy on both parts. In 827 Islamic armies from north Africa began the conquest of Byzantine Sicily.[38] It was also in war the advances of Arabic science contributed. Using the "wootz" process for producing steel developed by Indian smiths, Arabian centers in Damascus and Toledo made the finest swords of antiquity.[39] Wootz was produced by a crucible process using black magnetite ore, bamboo charcoal, and the leaves of certain carbonaceous plants. Smelting in charcoal fire of hot blast using high temperatures produced a regulus or cake of metal. This high quality steel (1–2% carbon) had a specific pattern after forging called "damask". Called "Mohammed's Ladder", it is the result of the separation of coarse and fine iron carbide particles.[40]

Islamic scholars, scientists and engineers owed more than skills in iron metallurgy to the Indian Hindu thinkers. Al-Bírúní wrote that India was famous for astronomy, and its attendant knowledge of mathematics, which was contained in books called the Siddhāntas ("Sindhind" in Arabic)[41] written after 400 A.D. These writers had used the gnomon and clepsydra for measuring angles and time. The first great algebraist was Āryabhata I who lived about 499. He developed a very accurate value for pi of 3.1416 (3 177/1250). Brahmagupta (ca. 628) followed, developing the theory of diagonals and areas of cyclic quadrilaterals (four-sided figures with their angles on the circumference of a circle). Mahāvíva (ca. 850) dealt with right triangles with whole-number sides. Bhāskara (c. 1114–?) summarized Hindu mathematical thought in his work on astronomy *Siddhānta-Siromani*.[42] Indian contributions to mathematics were principally in (1) solutions of indeterminate equations, (2) arithmetical and algebraic notation as well as decimals, and (3) trigonometry. Buddhism was responsible for the spread of Hindu knowledge to China, while Islamic conquest of eastern lands brought this knowledge first to their attention.

Al-Battāni (Albategnius), who lived from 858 to 929, developed tangent and cotangent functions from the Indian practice of the gnomonic angle measurement (*umbra versa* and *umbra extensa*).

Muhammed ibn Mūsā al-Khuwārizmi was the most influential of medieval Islamic mathematicians. His work synthesized Hindu and Greek mathematics with that of Islam, introducing the "arabic" (Hindu) numeral system to the west. Alforism and algebra derive from his work. He wrote extensively on solutions of linear and quadratic equations. An example is a geometrical solution of the quadratic $x^2 + 10x = 39$, by the "completion of the square" originated by al-Khuwārizmi and reproduced by 'Umar Khayyàm in his *Algebra*.[43]

Let us imagine that x^2 is the square whose side x is to be found. If we suppose the square to be drawn and then increased by the addition of four rectangles, each of area 2.5 times x, along its four sides, then the whole area is $x^2 + 10x$. To make a new and larger square we must "complete" this square by the addition of the four small squares at each of the four corners of the original x^2. Each small square has an area of (2.5 × 2.5), i.e., 6.25, and their total area is 25. Our final square is thus of area $x^2 + 10x + 25$. But $x^2 + 10x = 39$, hence $x^2 + 10x + 25$ must equal 39 + 25, or 64. So 64 is also the final square. The side of this final square is clearly 8. By subtracting from both ends of the side (8) of this square the two lengths of 2.5, we obtain the required value of x, which is to be 3.

Mathematicians will know of a further solution, which is $x = -13$, but the Arabic scholars dealt only with the positive answer. Al-Khuwārizmi also wrote on Ptolemy.

The East of Islam also produced the three greatest Arab minds in al-Bírúní (973–1048), Ibn Sínā (Avicenna) (980–1037) and Ibn al-Haitham (965–1039). Al-Bírúní was one of the greatest scientists in history, clearly a match of the likes of Archimedes or Aristotle. He wrote on metals, gems, hydrostatics, conic sections, astronomy and geography.[44] Ibn Sínā so dominated medicine that his texts were used into the 20th century. In many ways his work may have hindered further study as it was so definitive. He studied the transmission of light, correctly deducing that it was composed of particles traveling at infinite speed. His writings include treatises on heat, force, motion, and chemistry. Ibn-Haitham, better known as Alhazen, was a true physicist whose studies of optics and the eye were as advanced in their area as was Ibn-Sínā's in his.

Alhazen anticipated Fermat with the observation "that light, passing through a medium, takes the easiest and quickest route", e.g., the Principle of Least Time. He also understood and described the Principle of Inertia in mechanics. A colleague, al-Khāziní (ca. 1122), wrote *The Book of the Balance of Wisdom* on mechanics and hydrostatics. It also contained one of the first discussions of the theory of gravity as a universal force. Eastern Arab science reached its zenith with these men and declined thereafter, although men like Ulūgh Beg (ca. 1420), a grandson of Tamerlane (Timū Lang), the destroyer of Baghdad and the Tigris-Euphrates irrigation systems, established the finest astronomical observatory of that time in Samarkand.

By the 1100s the Conquests were over and the waning of the empire began. Cordoba by now had street lights[45] and Grenada had the great and beautiful fortress, Alhambra. By the 12th century al-Andalus was crumbling due to Christian pressure from the northern kingdoms of Aragon and Castile, as well as Portugal. Disputes between African and Spanish rulers further accelerated the fall of Moorish Spain. The torch that had passed to it fell from its hands by 1492 with the fall of Grenada. Its culture and learning passed away but the knowledge it had fostered and advanced returned to a Western world freed from the barbarism it had brought in its conflict with ancient Rome.

It is never wise to speculate too broadly on the reasons a culture seems to be less creative in a specific area than another. Byzantium was heir and keeper of classical science and philosophy. Why the contributions of this civilization seem less impressive than that of the Islamic world of the Middle Ages may be in large part due to the inherited baggage of that great heritage. A major point in this regard is the Hellenistic disdain for experimentation and the actual objectification of a theoretical concept be it in science or engineering. Constantinople was Roman but its culture and language was Greek. As heir to the Greco/Roman cultural legacy, it may have been too much to expect that they could easily put it aside and become as adept in practice as they were in theory.

Byzantine rulers from Justinian on did not help. Dissident sects, Christian or pagan, were persecuted and their scholars fled. Nestorian Christians, Jews, Monophysites, and philosophers first migrated to Iran then Damascus and Baghdad. Centers like Jundisshapur (Iran), Basra, Baghdad, and Cordoba were built on much of the knowledge and skills these men brought with them out of Christendom.[46] Intellectual activity became the province of the church in Byzantine. Indifference to science was more the rule than not. Following St. Augustine's injunction that this life (his was lived within a crumbling western Roman world) was a "veil of tears" and it was more important to concern oneself with the next, the Byzantines made religion central to their daily life and scholarship. Even so, the blame cannot be placed with one bishop or one emperor (such as Diocletian who froze trades and occupation to help stabilize the 4th century Empire). The lack of emphasis on classical science and engineering began before Byzantium had to go it alone, as it were, without the Latin west.

Nonetheless, the Arab mind blazed with a brilliance second only to the Greeks of the Classical Period. If the Greek was the father then the Arab was the foster-father of science and engineering.[47] For over two centuries Arabs and their subjects wrote and built on the Classical foundation they eagerly embraced. They equalled and surpassed the Romans in hydraulic engineering; building and fortifications were borrowed without much innovation beyond adiabatic cooling, pointed arches, and machicolated walls (cantilevered wall parapets). They cared little for roads, never adapting the Roman system to their use. In fine technology and power they surpassed their predecessors. Sciences such as chemistry, mathematics, astronomy and even that of agriculture are more derivative of Arabic knowledge than we admit.

As concerned with religion as the Byzantines, the Arabs and Islam in general were able to compartmentalize the secular and sacred to the advantage of practical life. This was best epitomized by the great philosopher Ibn-Rushd (Averroes) of Cordoba (1126–1198) who developed Aristotelian thought such that scientific and religious truth was sharply divided.[48] Even at this, the tension of these philosophical approaches led other Islamic scholars like al-Ghazzâli (1058–1111) to argue that science and philosophy undermined religion and led to loss of belief. His views seem to have prevailed, as by the 12th and 13th centuries Islamic learning was in decline. Like Rome much of this decline was due to external as well as internal forces.

Byzantium and the Caliphate fell before the Seljuk Turks. We first meet them under Alp Arslam at the fateful Byzantine defeat at Manzikert in 1071. They were muslim, a converted mongoloid race. They were orthodox Sunnites as opposed to the Shiites of Palestine, Syria and Egypt. These two strains of Islam are as political a division as religious, tracing back to the feuds over succession to the Prophet's place in the 7th century cities of Mecca and Medina.[49]

The Turks were also barbarians, just as the Germanic invaders of the 5th and 6th centuries were. They were users and not builders. Onto their realm they grafted Arabic learning and Byzantine pomp. They were more successful in assimilating the latter. They were, in the tradition of Islam, militarists. This they had to be to withstand new barbarians out of Asia under Jenghis Khan. Pouring out of the Steppes, his horsemen destroyed Islamic kingdoms in Iran, Turkestan, and Iraq, leaving only wastes where cities and irrigated fields once stood.[50] That the Seljuks were able to blunt this tide is to their credit.

NOTES

1. Africa was occupied by the Vandals who took Carthage in 439 A.D. and sacked Rome in 455 A.D. The Vandals were sailors who raided Mediterranean shipping until Belisarius and his Byzantines re-established Roman control in the 6th century. They were completely absorbed in the Arab conquests that came in the following century.
2. Vandal sea power set up a state in Africa A.D. 429 which fell apart under Byzantine attack in 533.
   Maiteus, J. The Vandals: myths and facts about a Germanic tribe of the first half of the first Millennium A.D. *Archaeological Approaches to Cultural Identity.* Sherman, S. J., Ed., University of Hyman (1989) p. 58.
3. **Hill, D.** *A History of Engineering in Classical and Medieval Times.* Open Court (1984), p. 32.
4. **Sandström, G. E.** *Man the Builder.* McGraw-Hill (1970), p. 136.
5. **Kirby, R. S., Thington, S. W., Darling, A. D., and Kilgour, F. G.** *Engineering in History,* McGraw-Hill (1956), p. 91.
6. **Sandström,** op. cit.
7. **Sherrard, P.** *Byzantium.* Translated by Richard Atwater. Michigan (1961).
8. **Procopius.** *Secret History.* Time (1966), p. 141.
9. **Hetherington, Paul.** *Byzantium, City of Gold, City of Faith.* Golden (1983), p. 21.
10. **DeCamp, L. S.** *The Ancient Engineers.* Ballantine (1963), p. 286.
11. **Casson, L.** *Illustrated History of Ships and Boats.* Doubleday (1964), p. 42. Ca. 600 A.D. saltpeter added to crude oil, sulphur mixture to make a spontaneous combustible fluid.
12. **Forbes, R. J.** *Man the Maker.* Henry Schuman (1950), p. 102.
13. Ibid., p. 98.
14. Ibid., p. 92.
15. **Dobrovolny, J. S.** *A General Outline of Engineering History and Western Civilization.* Stipes (1958), p. 67.
16. **Burke, J.** *The Day the Universe Changed.* Little, Brown (1985), p. 40.
17. **Dobrovolny,** op. cit., p. 19.
18. **Forbes,** op. cit., p. 96.
19. **Hill,** op. cit., p. 94.

20. **Philon of Byzantium.** *Pneumatics,* LXIII (ed. Cana de Vaux), p. 205.
21. **Vitruvius.** *De Architectura.* Loeb Classics (1970), p. 307.
22. **Hill,** op. cit., p. 134.
23. **Smith, N. A. F.** *Man and Water.* Peter Davies (1971), p. 20.
24. **Collett, J.** Water Powered Lifting Devices. Unpublished paper prepared for the Land and Water Development Division, FAO. (1980).
25. **Russell, G. E.** *Hydraulics.* Henry Holt (1937), pp. 354–356.
26. **Hill,** op. cit., p. 165.
27. Ibid., p. 54.
28. Ibid., p. 55.
29. Ibid., p. 57.
30. **Burke,** op. cit., p. 37.
31. **Brett, M.** *The Moors.* Golden (1985), p. 12.
32. **Hill,** op. cit., p. 21.
33. **Burke,** op. cit., p. 37.
34. **Brett,** op. cit., p. 45.
35. **Burke,** op. cit., p. 38.
36. **Hill,** op. cit., p. 58.
37. **Smith, N.** *A History of Dams.* Peter Davies (1971), p. 1.
38. **Brett,** op. cit., p. 16.
39. **Forbes, R. J.** *Studies in Ancient Technology.* Vol. 2, Brill (1965), pp. 207–208.
40. **Maugh, T. H., II.** A metallurgical tale of irony. *Sci. News,* **215**: 153 (1981).
41. **Winter, H. J. J.** *Eastern Science.* John Murray (1952), p. 37.
42. Ibid., p. 46.
43. **Kasir, D. S.** *The Algebra of Omar Khayyām.* (1931).
44. **Winter,** op. cit., p. 71.
45. **Attenborough, D.** *The First Eden.* Little, Brown (1987), p. 137.
46. **DeCamp, L. S.** *The Ancient Engineers.* Ballantine (1963), p. 303.
47. **Wells, H.G.** *The Outline of History.* Doubleday, Vol. 1 (1971), p. 527.
48. **Wells,** op. cit., Vol. 2, p. 528.
49. Ibid., p. 559.
50. **DeCamp,** op. cit., p. 309.

7 THE MIDDLE AGES — MASTERS OF STONE

Charlemagne attempted to bring western Europe a semblance of political order and education, the first since the collapse of the West. He failed due to the inherent weakness of Frankish royal succession which prevented his empire from remaining intact after his death in 814 and the strength of marauding Scandinavian tribes we call "Vikings". No place in Europe and even the near East that fronted a sea or a navigable river was secure from these new barbarians who, instead of riding horses, came on the backs of dragons.

These "dragons" were the Viking longships called "drakuschiffen" or dragonships because of their tall prows and sterns, often carved with fierce replicas of these mythic creatures (Figure 7-1).[1]

VIKING SHIPS

The ship design of the north European waters in the "Dark Ages" was a radical departure from that seen in the Mediterranean. The Nydam boat was an early step in this separate tradition called the "clinker-built ship". Clinker-built refers to the practice of over-lapping the hull strakes or planking rather than edge-joining the strakes. By 600 A.D. vessels built in Norway had true keels rather than center planks. Further, they were designed and built to carry sail by 800 A.D. The Gokstad boat was typical of Viking "Langschiffen" or longships. Its length was 23.3 meters and its beam was 5.2 meters. Sixteen lapped and riveted oak planks per side with a keel comprised the hull. The genius of the northern ship builders was in the joining of a strong hull to a frame rather than fastening individual planks to a frame as did the carvel hull designers of the Mediterranean. An example of a Saxon craft (ca. 700 A.D.) is the magnificent Sutton Hoo ship, a burial of an East Anglian king. Its dimensions were 24 meters in length, 4.27 meters in breadth and a draft of 61 centimeters. It was clinker-built of 3 meters long, 2.5 centimeters thick oak planks with 26 ribs, tholes or gunwales. The Graveney Boat (900 A.D.) was of the clinker-built tradition. Found in the Kent River, it was a smaller merchantman with a keel of 5.5 meters. The ribs were of unequal length — some spanned the keel and then had extension pieces that ran up to the gunwales — some ribs only ran to the turn of the bilge called "intermediate frames", copper nails used as fasteners and lead rings were sewn to sail to take the brails (ropes that ran up from the deck through the rings, over the yard and down again). Stone was used as anchors and as ballast.

VIKING SHIPS — DESIGN CONTRIBUTIONS TO THE PRESENT

Two principal types of strain act upon a ship. These are (1) a vertical shear force and (2) a horizontal bending moment. Shear force (1) is produced by two other forces — gravitation (g) and buoyancy (B). The former pulls the ship down and the latter thrusts it up. Where the two forces meet is where the ship floats, i.e., waterline. The more mass, the more g. The shear force is directional along vertical members of a ship and is a factor in differential loading of the hull by the distribution of the two forces of g and B. Stresses induced by extreme imbalance of g and B can cause hulls to fracture (Figure 7-2).

The bending moment (2) is the limit of flexure or pliability of a hull. In the phenomena of "hogging" (the hull bends upward) and "sagging" (the hull bends downward), hulls flex by design and by accident. This is due in large part to the forces that produce the shear force but act along the horizontal axis.

FIGURE 7-1.   Viking ship (Library of Congress).

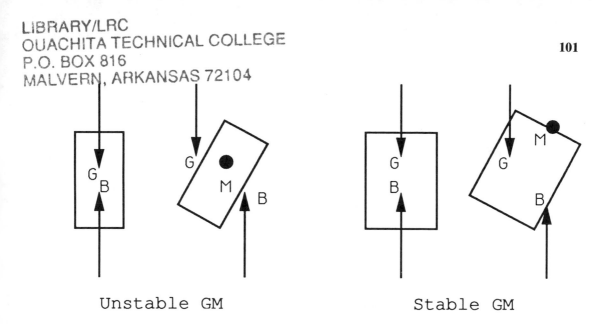

Unstable GM   Stable GM

FIGURE 7-2.   Diagram of buoyancy and gravitation forces acting on a ship hull.

In ancient ships like the trireme and the Pharoah's boats this horizontal axis was the longitudinal axis due to the obvious factor of the designed imbalance of length to breadth ratios of 3:1 to 7:1. A hull can be too rigid as well as too flexible. Either extreme can be disastrous to a ship. The ancients always sought to reach the point in hull strength where the bending moment was optimum. To create a strong frame and hull with a low value for the bending moment might cause a disparity in the other forces that comprise the shear force. For example, a massive planked hull with strong frames might increase g to a point the vessel could carry little mass but itself. The fully rigged Viking ship (Figure 7-1) was one of the strongest vessels for its size and purpose built in ancient times. It was one of the high points in ship design until the Clipper ship of the 19th century. Their construction provided a relatively light hull with great longitudinal strength in the shell firmly held by a complete interior frame. With the continuous, single-piece external keel, scarfed to stern and stern posts, the Norse boats provided the strongest craft with strength-to-weight ratios seldom exceeded today.

By 1000 the Vikings had shot their bolt. The Danish kings were driven from England by Alfred ("the Great") by 878 and their tide receded across Europe. In France they settled into Brittainy as vassals of the king and Duke of Normandy or William the Conqueror. He defeated Harold, the Anglo-Saxon king, in 1066 and ruled the English realm Alfred had restored. The Normans used armored heavy cavalry and military engineering to subdue England. The heavy cavalryman evolved into the medieval knight and the Norman motte and bailey castle became the medieval fortress.[2]

MEDIEVAL FORTRESS ENGINEERING — THE INGENIATOR
Motte-Bailey Castle Design

The medieval fortress ("castle") has its origins in the enceinte of Constantinople originally begun by Theodosius in 413 A.D. and completed, after destruction by earthquake in 447, during the reign of Justinian (483–565). This fortress wall design and over 700 fortifications built or reconstructed by Justinian became the models for the Crusaders, who came to the Middle East.

The medieval fortress was built about a central "donjon" or keep. This tower structure was self sufficient in concept and was first utilized by the Normans in the 11th century, in particular, William the Conqueror. The "motte and bailey" castle became the standard Medieval design. The motte was a man-made mound surrounded by a trench, and on it was erected a tower of timber. The bailey was a kidney-shaped enclosure at the bottom of the

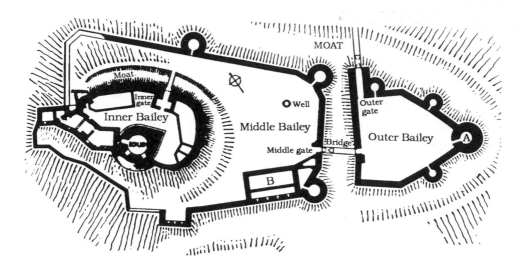

FIGURE 7-3.   Chateau Gaillard (1197).

mound intended for horses and cattle. The works were surrounded by a timber palisade. The motte measured 30 to 90 meters in diameter, and its height 3 to 30 meters. With enough men, the earthworks could be completed in a week. William was reputed to have utilized prefabricated timber towers in his landing to provide fortifications for his assault force of Normans.

The motte and bailey castle replaced the Anglo-Saxon fortified manor, which was palisaded as well. The Normans utilized the strategem of erecting a motte in the new territory and basing tactical operations from it. They used this device successfully in Normandy and France, as well as in the conquest of England. "Motte" means height in Medieval French.

The supervision of erection of the motte was by Norman archers but a specialist was retained for timbering, termed a *carpentarii*. Three names that come down to us from the Domesday Book are Stefanus, Durabus, and Raynerius. These men and others of their specialty erected 40 motte and bailey castles in the first 20 years of the Norman Conquest. The first masonry castle was built by Gondolf, Bishop of Rochester, in 1087. His greatest achievement was the design and supervision of the construction of the Tower of London in 1097. This stone keep, 33×57 meters, replaced the older motte and bailey.

The Tower's "bailey" was actually composed on two sides by the original Roman wall, and earthworks on the other two sides. A defensive tower, the Bell Tower, was built in the southwest corner (1200). Henry III converted the Tower into a concentric fortification with a curtain wall and towers on three sides with a widened moat (1225–1260). He strengthened the Roman wall as well. By 1300 the old moat was filled in and a newer wider one was dug. The west curtain was built of stone. The shore was filled in and works built on the fill to include the "Traitor's Gate" as a sea gate on the Thames.

The influence of medieval social changes was reflected in the role and design of the castle. With a weakening in central authority, i.e., a strong king such as Charlemagne or William, the landed aristocracy became semi-autonomous of the king, holding large estates with wide powers of justice and administration ceded to them.[3] The castle was thus a caput of a lordship or chateaux of a siegneur[4] instead of simply a military stronghold.

With this change in role, the design of castles changed as well. The "shell keep" appears in the later 12th century. This construction was a hollow, bearing-wall structure atop the motte rather than the monolithic-type keep of earlier times. This wall design was applied to the bailey with the use of a "curtain" of stone between towers or fortified wall elements.

Chateau Gaillard, built in 1197 by Richard the Lionhearted (Figure 7-3), reflects this

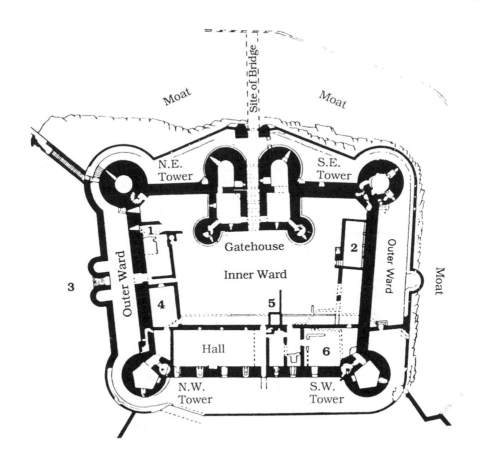

FIGURE 7-4. Concentric castle plan. (Redrawn from Sandström, 1970.)

change in design, together with the influence of Eastern Empire design ideas as seen in Byzantine fortifications such as Justinian's Wall at Constantinople. At Arundal and Windsor in England the shell keeps were rectangular in plan, with walls so thick that chambers and wall passages were built in their upper parts. No vaulting was used, only a central partition rising the full height of the tower to help support the roof, in two spans with a central gutter. Dover Castle was the last English keep built as a square tower (1180–1190).

The military weakness of the square tower design became evident as the four salient angles offered security to siege parties as the sappers were out of reach by direct vertical fire from defenders. The sapping for medieval castles was less difficult than one would suppose, as the construction of the shell and curtain walls consisted of ashlar filled with rubble. Once the shell was pierced, the rubble fill could easily be pried out. To combat this weakness, the round tower and, in the case of English designs, the polygonal tower were used. The polygonal design had the advantages of the round tower but not its inherent strength.

Concentric Castle Design

The evolution of the Norman castle followed the course of the subjugation of Wales. The Welsh Marches became a tension zone that the Norman Lords never quite controlled as with the rest of England after the Conquest. Welsh resurgence in the 13th century precipitated new castle building by Edward I, beginning in 1276. Master James of St. George, Edward's master military architect, utilized a new type of castle design, the concentric castle, in which the higher inner walls permitted archers to fire simultaneously with those on the lower curtain wall (Figure 7-4). This effectively doubled the defense's fire power. The Welsh

TABLE 7-1
Various Measures Used in
Medieval Construction

Roman foot	295 mm
English foot	305 mm
Royal foot	325 mm
Teuton foot	333 mm

castles were inspired by the symmetrical fortifications in the Crusader Kingdoms of Jerusalem and Sicily. Together with the replacement of keep with retangular or polygonal curtain walls, and symmetrically placed towers (round or polygonal) placed one inside the other, a powerful gate-house with a barbican and postern was built. Harlech and Rhuddlan were concentric, as was the later Beaumaris Castle of North Wales (1297).

Master James originally built castles in Savoy for Edward's cousin, Count Phillip of Savoy. He added St. George to his name after moving to England. For his work for Edward, Master James was paid a stipend of three shillings a day plus a pension of one shilling, one penny a day for his wife in the event of his death. We see among his key staff a Master Bertram de Saltu, the "ingeniator". Master James, considered the greatest ingeniator and military architect of the Middle Ages, died in 1309. Beaumaris is considered his masterpiece of medieval military engineering, with upwards of 300 firing positions — moats and complicated access routes. An enemy that assaulted Beaumaris faced a new "killing ground" with each component of the castle's defenses from outer gate to interior sections. An assault on Beaumaris represented 14 separate battles of intense magnitude. Of further interest is the small garrison necessary for the defense of such fortresses. Caernavron, on a scale of that at Beaumaris, had 20 men-at-arms and 80 bowmen. Harlech was held by 10 men-at-arms and 30 bowmen. By 1325, the death knell of these formidable constructions was sounded with the first use of cannon in Britain. It was over a century before serious siege weapons were cast, with the reduction of Friesack Castle by Elector Frederick I by means of cannon in 1414. In England, the Earl of Warwick reduced Bamborough Castle in one week in 1464.

ENGLISH FORTRESS BUILDERS

Gondolf, Bishop of Rochester (1097), built the Tower of London. Master James of St. George, Ingeniator of Edward the First (1272–1307), built Aberystwyth, Flint, Rhuddan, Conway, Caernarvon, Harlech, and Beaumaris. Viollet-le-Duc, the French architect/historian, discussed the fortress and its design in his six-volume work on French architecture in the Middle Ages.[5] In the early Middle Ages attack was the prevailing tactic, but as the period drew on developments in castles and armor, it brought defense to the forefront. Towns were walled and fortified. Feudal knights were ultimately encased in steel shells. Viollet-le-Duc says the fortress did *not* evolve from the Roman *Castrum*, but rather from palisaded "fortress"-homes. By 1000 A.D. these castles were of stone. The Middle Ages "Ingeniator" became a master of the art of masonry.

In cathedral building, new vaulting techniques, beginning with the 11th Century, led to the development of the great Gothic cathedrals. The Romanesque vault was, in early phases a rounded barrel structure, later becoming a pointed barrel vault. There was no standard measure system in the Middle Ages, as evidenced by Table 7-1.

Domed construction was rare in large buildings, except in the Byzantine East. Two rare western examples were (1) the palace chapel at Aix-la-Chapelle (796–804 A.D.) and (2) the Baptistery at Florence.

In vaulted construction great thrusts must be carried by the side walls; typical wall thicknesses were 2.5 meters. From an engineering and economic standpoint these side walls

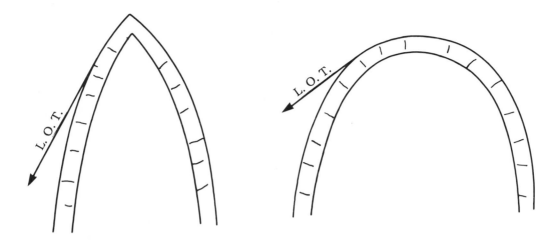

FIGURE 7-5. *Left:* Point Arch and Line of Thrust (LOT). *Right:* Semi-circular Arch and Line of Thrust (LOT). Arch forms taken from medieval drawings.

were somewhat wasteful of material. Some progress in lightening the side walls was made by use of arched recesses, but only with the development of the buttress as a device, in France, did the true Gothic style develop to its height in 1155–1220 A.D. The buttress works because the line-of-thrust generally leaves the arch below the quarter span. The cathedrals were built in amazingly short periods — 25 years at Chartres and even less at Laon.

The structural devices used by Gothic builders to cover maximum space with a minimum quantity of material and reach over 47 meters in height were (Beauvais):

1.   Distinction between bearing pillars and non-bearing walls serving merely for enclosure (i.e., windows)
2.   General application of the pointed arch
3.   Distinction between vault-supporting ribs and intermediate panels (14 cm thick panels were common) of lighter weight, i.e., a frame of stone ribs
4.   Perfection of a system of buttresses and "flying buttresses" to absorb vault thrust

The pointed arch is statically efficient. It has greater conformity with the line of thrust than the semi-circular arch, which deviates from the line of thrust at the quarter span (Figure 7-5).

The formulas, mathematical and geometrical rules of Gothic construction were handed down from master to master. These were concerned mostly with form and composition — six centuries would pass before a design of a building element would be made on the basis of structural analysis.

The Gothic Cathedral builders "Rector Fabricae" and "Ingeniators" were both engineers and architects (Table 7-2). Villard de Honnecourt (also called William) was one of these masters who traveled, alone or with a team of skilled workmen, from job site to job site, sometimes over long distances. Honnecourt's notebook gives us details of the Gothic vaults, truss bridges, and machines these men used.

THE PERSONAL RECORD OF VILLARD DE HONNECOURT (ACTIVE CA. 1230)

During the French Revolution a small (6 × 9) notebook of 30-odd vellum pages stitched into a thick, rough, overlapping leather cover was found at the famous Abbey of St. Germain-

TABLE 7-2
Names of Gothic "Rector Fabricae" and
"Ingeniators"

Peter Parter (b. 1325) — Prague Cathedral
Michael Parter — at Strasbourg Cathedral (Ingeniator)
Erwan Von Steinbach — Strasbourg (Elevations service) (Rector Fabricae)
Villard de Honnecourt (France — Hungary)
Gerardus Lapicide — Cologne Cathedral
Alan of Walsingham — Ely Cathedral

de Près in Paris and deposited in the Bibliothèque Nationale.[6] De Honnecourt's notes cover surveying problems like how to measure a stream without crossing it. He showed a rough wood frame, the two sides of which could be directed toward an object on the far side of a stream, the aligned frame turned, and a point sighted in, from which the distance could be measured. Figure 7-6 illustrates a page from de Honnecourt's notebook. The illustration shows a machine that trims the heads off underwater piling.

This interest in cofferdams and piling harks back to the Roman times, for the many bridges and highways of that day had been destroyed or had fallen into disrepair. Harbors had silted in. Engineers of the Middle Ages began the rebuilding of river bridges and harbors. They used floating pile drivers and bucket dredges.

De Honnecourt notes a recipe for cement:

> Take lime and powdered pagan (Roman) tile in equal quantities. Add a little more of the latter until its color predominates. Moisten this cement with linseed oil, and with it you can make a vessel that will hold water.

Villard de Honnecourt had to understand all elements of construction from geometry and carpentry to cement. Artisan-architect-engineer in one man, Villard wrote in his notebooks on the topics of (1) mechanics, (2) geometry and trigonometry, (3) carpentry, (4) design, (5) ornamentation, (6) statuary design, and (7) furniture.[6]

The geometry for construction of the great naves of the 13th century were based on figures as shown below (Figure 7-7). *Opus ad triangulum* was used for calculating height and proportion for a Gothic cathedral based on the equilateral triangle. This always produced a nave lower than one built using *opus ad quadratum*. The 1:2 ratio gave a wide nave and broad aisles, while *ad quadratum* gave a 1:3 relation between the nave and aisles.[8] Table 7-3 gives some nave heights of important Gothic cathedrals.

An interesting device developed by the builders of Westminster is the hammer beam truss illustrated in Figure 7-8.

The Christian attempts to recapture lands from the Muslim resulted in eight "crusades" from 1096 to 1291. In the process they sacked cities from Toledo (1085), Jerusalem (1099), to Constantinople (1204). While the "Latin Occupation" (1204–1261) of Constantinople resulted in looting of vast amounts of Byzantine treasures,[9] the capture of Toledo resulted in wealth of a different nature — books.

Arab scholarship of the 9th and 10th centuries resulted in the translation of a large amount of classical scientific works.[10] Together with Indian works and their own great contributions in algebra, chemistry and trigonometry, the Arab scholars and scientists filled libraries in great learning centers from Basra, Iraq to Toledo, Spain. With the conquests of Moorish cities such as Toledo the western "ingeniator" and builder such as Villard de Honnecourt had available works such as: Adelard of Bath's (1090–1150) translations of *Euclid* from arabic; his translation of the *Arithmetic* by al-Khuwarizmi; Robert of Chichester (6 years) translated "Algebra" (1145); Gerald of Cremona (1114–1187) translated Archimedes *On Measurement of One Circle* and Euclid's *Elements* into Latin; Eugenus of

Villard de Honnecourt: pile saw.

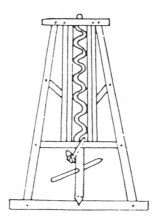

Villard de Honnecourt: screw-jack.

FIGURE 7-6.   Diagrams of pile saw and screw-jack (after de Honnecourt).

Palermo translated Ptolemy in 1160. By the 12th century this knowledge was available at universities in Paris, Chartres and Bologna. It is possible that men such as de Honnecourt studied these works directly rather than as journeymen on construction sites. Master masons in today's France often have advanced degrees in construction, as well as mathemathics.[11] Crosby concluded in his study of Saint-Denis that Piere de Montreuil (1239) based his reconstruction of the nave and transepts on an *ad quadratum* plan[12] or 1:2. The conclusion was that de Montreuil was aware of Arabic knowledge, as was de Honnecourt, which went beyond pure empiricism or mockish transcription.

ELY CATHEDRAL — ALAN OF WALSINGHAM

Ely Cathedral, located 65 miles north of London in the vicinity of "The Wash", is a former fen area drained by the work of Rennie and Vermuyden. In the early 1300s[13] the Norman tower of the Ely Cathedral transept crossing fell. The weak point in the Norman design was the tower at the crossing, particularly when founded on poorly understood or

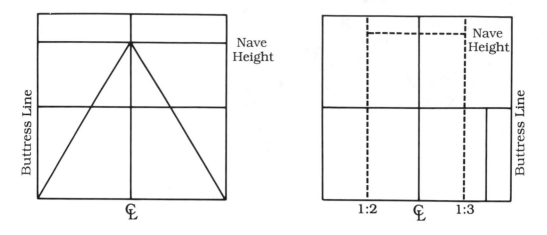

FIGURE 7-7.   *Left:* Opus ad triangulum. *Right:* Opus ad quadratum.

TABLE 7-3
Nave Height for Some Gothic
Cathedrals

Cathedral	Date	Nave height (meters)
Westminster	1245–1272	31.6
Chartres	1194–1225	34.4
Notre Dame	1163	35.0
Rheims	1211	38.1
Amiens	1230–1240	42.6
Palma	1355–1370	44.0
Beauvais	1272	47.8

known foundation such as the soils at Ely. One pier settled more than the other three, causing more stress on that pier and leading to collapse of the tower. Alan of Walsingham, an official of the cathedral and son of masons, undertook reconstruction but proposed and executed a design between 1323 and 1330 that broke with the 13th century discipline of right angles.[14]

Alan proposed an octagonal shape founded on the piers at the ends of the aisles of nave and transept. He redug the foundation of the piers to solid footing and used rock and sand fill with mortar to secure his foundation. On the piers he erected the wooden octagonal structure with wooden cantilever arches springing to a lantern ring. Eight to nineteen meter vertical posts >1 meter in diameter rose from the platform at the level of the lower ring. Two posts were only 16–17 meters, so Alan's carpenter consultant, William Herle, spliced the two with shorter sections to reach the required height. Compression and tension in the ring structure was ca. 27,200 kilograms. The timber was faced with mortar and covered with lead. The mortar (and plaster) protected the wood from chemicals in the sheathing. The timber octagon is turned through $22\frac{1}{2}$ degrees relative to the stone piers of the lower octagon.

The lantern cupola provides dramatic amounts of light into the crossing. Later workers placed a belfry on top of the octagon but this was removed in the 1700s, thus eliminating unnecessary dynamic forces. The entire octagon is a statically determinate structure and a little deviation in the settlement of one member has little effect on the relationship of the other vertical or horizontal components (Figue 7-9).

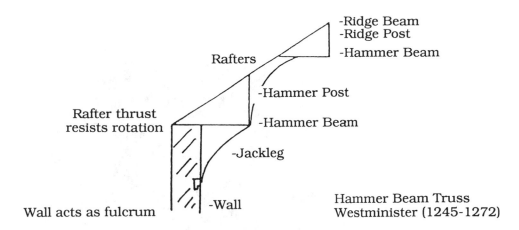

FIGURE 7-8. Hammer beam truss.

Alan later designed the Lady Chapel (1321–49) at Ely but is best remembered for the octagon. Alan (1) utilized wood as a principal material in a dramatic structural design, (2) utilized a new geometry in late Gothic design, and (3) did some of the first systematic foundation design in engineering. Alan of Walsingham was elected Bishop by the Ely chapter but an expatriate who had the Pope's ear at Avignon was elevated over Alan. Later Edward III, to aid in his French Wars, raised his treasurer, Simon, to fill the post after Alan was elected, again, upon its vacancy. By the time Simon became Archbishop of Canterbury, Alan was dead. His chapter had the last laugh as his monument has the sceptre and croxier of the Bishopry in Alan's carved hands.

During the Revolution, Cromwell's soldiers sacked Ely and destroyed all the medieval stained glass, as well as the statues in the church except for one — Alan of Walsingham.

ROADS AND BRIDGES

Most medieval roads were dusty or muddy tracks, depending on season. Roman roads were still in use here and there.[15] Surviving bridges were still important spans until the 12th century. Until this time most medieval spans were wooden trestles. Stone bridges were built over the Danube at Ratisbon (1135–1146); Elbe at Dresden; Main at Winzberg; Moldau at Prague; Thames at London; and Arno at Florence (Ponte Vecchio). Many types of soffit were used — semi-circular order, oval curves to flat segmental arches. Some bridges had rises of only $\frac{1}{5}$ of the span, as strongly arched bridges were a hinderance to pedestrians and animal-powered wagons.

The boldest bridge design of the Middle Ages was the great bridge over the Adda at Trezzo, in Northern Italy, which spanned a river with a single arch of 72 meters in length with nearly 21 meters rise (1370–1377). Due to its strategic importance it was destroyed in 1416. Today only the abutments remain, with small overhanging remnants of the vault. This vault was about 2.26 meters thick and 76 centimeters wide. It was twice the span of the longest Roman-designed bridge, only to be surpassed in modern times with the use of modern concrete and reinforced design.

MEDIEVAL THINKERS AND STRUCTURAL ANALYSIS

"Scholasticism", which aimed at the reconciliation of faith and reason, unwittingly promoted the development of mathematics and science. Roger Bacon, the great adversary of Scholasticism, was trained at Oxford and Paris and believed *"Experience, Experiment, and Mathematics"* to be the pillars of natural science. For the first time in 1000 years men discussed problems that were to lead to the development of modern "structure analysis" in

FIGURE 7-9.   Ely Cathedral. (Library of Congress)

engineering. The first to write on statics was Jordanus Nemorarius (ca. 11th or 12th centuries). This author goes back to the few then known fragments of Greek learning concerning the level (Euclid) and mechanics (Aristotle). In his treatise Jordanus develops the concept of the gravity dependence on the position of an object, e.g., he contemplates a weight moveable on an inclined plane, the one and only active component of the weight parallel to the track of the object. The flatter the inclination, the smaller the gravitational component. This notion leads to reflections on the rectilinear and angular lever. Jordanus, however, did *not* develop a correct notion of the static moment. His proof of a rectilinear lever of unequal arms implies the Principle of Virtual Displacements, e.g., '' a force able to lift a weight G to the height h is also able to lift a weight G/n to the height n × h''.

MAN AND MEDIEVAL MECHANICAL INNOVATION

Lynn White[16] has concluded that the post-Classical concern with the reduction of human and animal labor is tied to the development of prime movers other than these. Mumford goes on to assign this development to an ethos or view by Western men that work itself was not the degradation the Greek philosophers taught but a moral obligation of the Christian.[17] Bertrand Gille has presented documentary evidence to show the monastery, in order to achieve leisure for contemplation, went in heavily for power machinery and other labor-saving devices. The monastery became the model industrial village.[18] Gille's and Mumford's thesis is that the demand for technical innovation came from the culture, in this case the nonsecular monastic orders which flourished in medieval Europe.[19]

St. Benedict founded Monte Cassino in the Italy of the 6th century, about the same time as St. Augustine in Carthage was writing his treatise *The City of God*. Both men viewed the crumbling Roman works as something to be lived outside of both physically and intellectually. To do so required the founding of Benedictine monasteries which subscribed to Augustine's philosophy and Benedict's Rule of Life, which organized the day around spiritual contemplation and manual labor.

Add to this period a signal change in agriculture, where the farmer now had at his disposal a heavy, wheeled plow which used a sharp blade to cut the furrow, a share to slice under the sod, and a mold board to turn it over.[20] No two oxen could pull this machine. It required eight to pull it along the field, which was now plowed in long strips. Three-field rotation was now practiced and horses began to replace oxen. Productivity doubled in many cases and the resulting purchasing power fueled the medieval economy. The towns grew and the great churches were built.

To handle the increase of grain for the growing population, more mills were necessary. From the late Roman period the water-mill now became a fixture in medieval life. Every village had at least one for milling and larger mills were used to make beer and aid in fullery, tannery, iron forging, and mining. White tells us the horizontal-axle mill was in use by 1180 and windmills were prevalent from England to the Near East.[21] Gimpel, in *The Medieval Machine*,[22] calls the Middle Ages the "First Industrial Revolution".

Coupled with a changing climate which made Europe better suited for cereal agriculture, better diet, and labor-saving machines, and aside from the assorted Crusades and high infant mortality, the medieval world was an improvement over the dark days just after Rome's fall. Instead of a "wood and water" technology, perhaps we should term the Middle Ages the period of "water, wind, and the horse".

NOTES

1. **Almgren, B.,** *The Viking,* Crescent (1975), pp. 247–281.
2. **Bur, M.,** The social influence of the motte-and-bailey castle, *Sci. Am.,* **248**(5): 132-139 (1983).
3. **Kerr, N.,** Welch Castles, *Popular Archaeol.,* **5**(1): 3–8 (1983).
4. **Bur,** op. cit.
5. **Viollet-le-Duc, E. E.,** *Dictionnaire Raissoné de mobilier français de l'epoque Carlovingienne à la Renaissance,* (6 vols.). Gründ et Maguet (1914).
6. **Boure, T.,** *The Sketchbook of Villard de Honnecourt,* Indiana University Press (1959).
7. **Gimpel, J.,** *The Cathedral Builders,* Grove (1961), p. 109.
8. **Sandström, G. S.,** *Man The Builder,* McGraw-Hill (1970), pp. 114–115.
9. **Forman, W.,** *Byzantium,* Golden (1983), p. 57.
10. **Gimpel,** op. cit., p. 120.
11. Personal communication from Professor Vivian Paul, 1987.
12. **Crosby, S. M.,** *The Abbey of St.-Denis, 455-1122,* Yale (1942).
13. **Pevsner, N.,** *An Outline of European Architecture,* Penguin (1968), p. 138.
14. Ibid.
15. During the Middle Ages adequate safeguard against robbers was reflected in the "right-of-way" width.
16. **White, L., Jr.,** The flowering of medieval invention, *The Smithsonian Book of Invention* (1978), p. 64.
17. **Mumford, L.,** History: Neglected clue to technological change, *Technology and Culture,* Vol. II (1961), p. 235.
18. **Gille, B., Ed.,** *Histoire Générale des Techniques,* Partiemédiévale (1959).
19. **Gimpel,** op. cit., p. 15.
20. **White,** op. cit., p. 69.
21. **White,** op. cit., p. 67.
22. **Gimpel,** op. cit., p. 161.

8 RENAISSANCE

The complete rebirth of science and industry in the late Middle Ages suffered from the Hundred Years War in France and the "Black Death", bubonic plague, that killed up to one third of the population of Western Europe particularly in England, France, Spain, and Germany.[1]

In Italy, Rustidrollo of Pisa, imprisoned with Marco Polo at Genoa, wrote *The Travels of Marco Polo*. Polo told stories of fabulous wealth in Cathay and the Indies (East). They were not forgotten, nor were his assurances that China and its offshore islands were bounded on the east by an ocean that no doubt stretched unbroken to the coast of Europe. A Genoese merchant-sailor of the 15th century believed—Columbus—and sailed to cross that "unbroken sea".[2]

In Germany, Nicholas of Cusa (1401–1464) exposed the "Donation of Constantine" as a forgery and fraud. The brilliant scholar foreshadowed Copernicus and Newton with the idea that the earth revolved about the sun. He worked out the Gregorian calendar (Gregory XIII put into practice, 1582) and the concept of the infinite in mathematics.[3]

Back in Italy, Leonardo da Vinci began the 7000 pages of his *Notebooks* on flying machines, parachutes, helicopters, diving suits, tanks, and split-level cities. Also in Italy, recovery of artistic and literary treasures gave impetus to a growing Humanism, which ran counter to the philosophy of Scholasticism. This Humanistic outlook fostered new growth in the sciences and had a profound effect on engineering. Italy began its participation with the development of the lock, a hydraulic landmark of the 13th century.

LOCKS

In a post-medieval world where communication by roads was problematic in the best of times, the rivers presented avenues for travel and transport if they could be navigated. The greatest obstacle to the navigation of rivers was their slope. In earlier times goods or even boats, if they were light enough, were carried around obstacles such as rapids or bars. Another obstacle to the use or improvement of river transport was the political situation of medieval times. Without improvement in the latter, no progress could be made on the technical problems of navigation. In 1167 Milan and other northern Italian cities organized the Lombard League.[4] Within the area under its control these cities could plan and execute hydraulic projects. The canals in this area were the type built on a descending slope, which required a device to impound water to raise or lower barges and boats, e.g., the lock.

The lock has been considered the greatest single contribution to hydraulics ever made.[5] In the Lombardy canals the lock is mentioned as a "conche" or single barrier wier design. These single gate locks were difficult to use and were replaced with multiple gate locks subdividing the stream's elevation, thus reducing fall between barriers. This scheme was first described in a document from the Milan archives, dated 1445:

> If along a river...we wish to conduct boats [the accompanying text states], when due to little water and an incline it might be impossible to navigate, it is necessary to determine the fall.... Let us suppose that the first part of the river has a drop of 30 piede; construct at that point a high door in the manner of a portcullis...with windlasses to raise it, and in this manner lay off the entire length of the river and all its falls with such doors. After the boat enters, and the door is closed, the boat will soon rise...and will be able to enter the second chamber...and so step by step you will be able to take the boat to wherever you wish. Should you desire to return down, by opening each door, the boat with the water will be led to the next door, and so from one to the other it will be possible to return to the sea. All boats should be made with flat bottoms so that they will float on little water.[6]

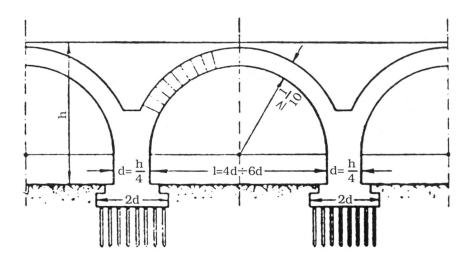

FIGURE 8-1.   Arch bridge, Alberti.

Leon Battista Alberti (1404–1472) took the design of the lock to the type that we associate with the device today. In *De re aedifactoria*, Alberti describes a hinged vertical gate lock that opened horizontally.[7] Beyond the lock, Alberti writes of a whole program of items for engineers — "engines, mills, channels, sea dikes, fortresses, rock cuts, etc".[8]

Empirical rules to determine dimensions become the germ of "engineering science". Alberti, in *De re aedifactoria* (1485), determined that: the Pier width of bridges (d) equals one quarter of the height of the bridge, Arch span (1) is equal to 4d divided by 6d, where d equals pier width. The voussoir thickness is equal to $^1/_{10}$ the span (Figure 8-1).

For pile foundations Alberti specifies a pile width equal to two times the width of the wall to be carried. Pile length is equal to one eighth the wall height. Pile diameter is equal to $^1/_{12}$ pile length.

The use of prime movers, other than man, was rare in the work of cranes and hoisting gear. The treadmill was the principal motive device for cranes and systems of gearing.

THE ORIGIN OF STATICS

Among the physical and mechanical problems examined by Renaissance scientists, two were of fundamental importance to engineering:

1. The Composition of Forces, force parallelogram, which is the basic problem of statics, and
2. The problem of Bending, elementary to the theory of the strength of materials

The problem of force components was tackled by Leonardo da Vinci, Stevin, Roberval and others before it was solved by Varignon and Newton. The problem of bending engaged Mariotte, Hooke, Jakob Bernoulli, Leibnitz, Parent and others. The final solution had to await the 18th century with Coulomb and Navier. To successfully deal with statics it was necessary to define the concept of "force" as directional or vectorial. This conceptualization is a considerable effort of abstraction. At the beginning of the 16th century little distinction was made between "force" and weight ("gravitas") as a special form of force. For example, Jordanus had still but a vague notion of nonvertical forces. Leonardo da Vinci (1452–1519) dealt with center of gravity, principle of the inclined plane, essence of "force" arriving at a fundamental conception of the statical moment (force times the distance from the fulcrum).

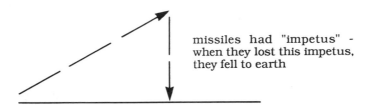

missiles had "impetus" -
when they lost this impetus,
they fell to earth

FIGURE 8-2.   Medieval notion of trajectory and motion.

The Dutchman Simon Stevin (1548–1620) dealt with center of gravity, level, and the inclined plane. He discusses the composition and resolution of forces in the case of a weight suspended on two strings. In *Mathematicorum Hypomnemata de Statica,* he makes correct use of the parallelogram of forces and the notion of the static moment. Italy produced Galileo Galilei (1564–1642), one of the greatest scientists and engineers since Archimedes, particularly in the study of mechanics.

He was the first to use the term ''moment'' meaning the ''effect of force''. He applies the term to mass times velocity and to distance from a fulcrum, reasoning correctly that in either case the same mass exerts a greater or smaller effect, according to velocity in one case, or distance from a fulcrum in the other. He founded the science of the strength of materials. He begins the cantilever beam equaling moments (of load) with the resultant of tensile stresses in the beam's cross-section with an axis of rotation at the lower edge of the embedded cross-section. Thus, he arrives at the correct conclusion that the bending strength of a rectangular beam is directly proportional to its width, and proportional to the square of its height (h). He errs by not introducing elasticity, realized by Hooke a half-century later, which is tensile strength vs. bending strength only.

Galileo's Moment of Resistance to Bending equals $bh^2/2$ where b = width, and h = height of beam, rather than the correct value $bh^2/6$. His real achievement is in the correct formulation of the problem. Galileo proved that the independence of the pendulum period of mass depends only on length, e.g., $T = 2\pi/1/g$, where 10.1 m/s (e.g., 2T (s) = 1 m in a ''grandfather'' clock).

MILITARY ENGINEERING

Fortress engineering, in addition to hydraulic engineering, continued to be that sphere of engineering in which theory and practice, architecture and mathematics, were most interwoven. Military engineers had to have a more profound knowledge of mathematics, being concerned with ballistics.

The term ballistics comes from the Greek word ''ballein'' meaning ''to throw''. Ballistics was the province of the military engineer for centuries. Mathematics of motion directly related to the study of ballistics. Medieval artillery included:

The *euthytonon:* bow with mount, fired large arrows or small spears.
The *palintonon:* fired stones along curved trajectory.
The *catapult:* rigid arms, twisted skeins for torsion; shot spears or stones.
The *trebuchet:* unequal arm with weight for mass to create ''teeter-totter'' effect, range of trebuchet, catapult was 100 meters.

To medieval science missiles had ''impetus'', which they lost and then fell to earth (Figure 8-2). The first to study trajectory systematically was Niccolo Tartaglia (1500–1557), an Italian mathematician. He observed the flight of cannon balls which he concluded had a

vertical range for
equal velocities

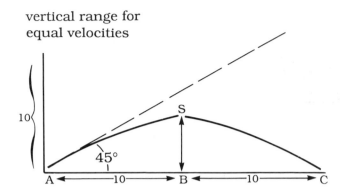

FIGURE 8-3. Tartaglia's concept of trajectory and range.

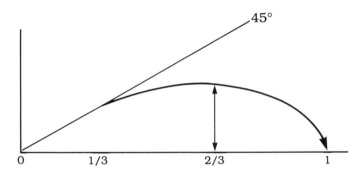

FIGURE 8-4. Trajectory as an ellipse, after Newton.

trajectory of three parts: a straight line, a curve and finally a straight line. He concluded the two "straight" sections might not be perfectly straight but so little curved they could be considered straight. The top of the curve he declared to be a quarter of a circle. Tartaglia did not understand air resistance, momentum and gravitation but intuitively knew elevation effected range. He set out to determine the elevation that yielded the greatest range. This turned out to be 45°. He invented the "gunner's quadrant" and formulated Tartaglia's Rule — Range AC = 2 AB_1 where AB_1 = 2(BS) and BS(also) = $^1/_2$ AB or $^1/_2$ BC (AB_1 is vertical range for equal velocities)(Figure 8-3).

Galileo rejected Tartaglia's three-part trajectory and opted for continuous parabolic curve. This is true only if the earth were flat but for ranges of ca. 8 kilometers, the differences between the parabola and the true curve are hard to determine or distinguish. Johannes Kepler (1571–1630) determined that the orbit of planets described an ellipse. In 1687 Newton (1642–1727) came to the conclusion that a trajectory was only part of an orbit (Figure 8-4). Actual trajectory is *part* of an ellipse (discounting air resistance).

C. Cranz determined the error in using a parabola or ellipse for range when the curvature of earth is considered, using 44° elevation and V = 820 m/s.

	Range	Peak Height
Calculated as Parabola	66,500 meters	16,538 meters
Calculated as Ellipse	68,958 meters	16,723 meters

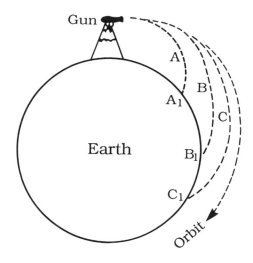

Mountain Gun

FIGURE 8-5. Newton's mountain gun.

The ellipse has the greater range — only 74.3 meters due to the wrong curve, 384 meters of range due to curvature of the earth. At ranges up to 16 kilometers a parabola is a good approximation. At ranges greater than 16 kilometers an ellipse must be used for calculations.

Consider a cannon on a high mountain (Figure 8-5). Its barrel is horizontal. When fired, the shell, due to momentum and gravitation will result in a curve A with impact A_1. If you increase the muzzle velocity, the trajectory B is shallower, likewise C is shallower still. As the curvature grows with velocity, the trajectory matches the curvature of the earth itself. You no longer have a trajectory but an orbit. This occurs at a velocity of 28,962 kph. Newton described this in his theory of artificial satellites. Actual trajectory was not a parabola or incomplete ellipse, lopsided.

Filippo Brunelleschi (1377–1446) was a Florentine who has been credited with the design of the "piece of engineering construction that gave birth to scientific engineering".[9] In 1407 he attended a conference of architects and engineers *(ingegneri)* on the method of erecting a dome for the crossing of the cathedral of Florence, Santa Maria del Fiore, which had been begun in 1296[10] (Figure 8-6). Brunelleschi, like other great builders of the Renaissance, was trained in a field other than engineering or architecture. He was first a goldsmith; Bramante, a painter; Raphael, a painter; and Michelangelo, a sculptor.[11]

Gille has Brunelleschi being an "architect" after his return from Rome in 1407.[12] The Pantheon dome made a great impression on him although he did not use the classical design in his dome. As we have seen, the cross-section of the dome is an arch of different shapes — semi-circular, segmental, elliptical, parabolic or pointed. The dome differs from the arch in that it is self-supporting as it is built up in steps of concentric rings (crowns). The arch is only self-supporting when the last keystone is in place. This is why the Pantheon has the great oculus at its center without danger of collapse. This oculus or hole can be open as in the Pantheon on covered with a structure called a "lanteen".

The thrusts in the crown of the dome are called "resultant crown thrusts", acting both tangentially and horizontally, all acting outwards. This tension must be restrained by the masonry or counteracted by some form of bands acting like barrel-hoops. Prior to Brunelleschi domes had been built as a single shell. In his dome, begun in 1420, he adopted a two-shell design (Figure 8-7) built on an octagonal plan. The outer dome was to protect the inner or main dome.[13] The two were cross-connected across the inter-space, yielding a thicker, stiffer structure.

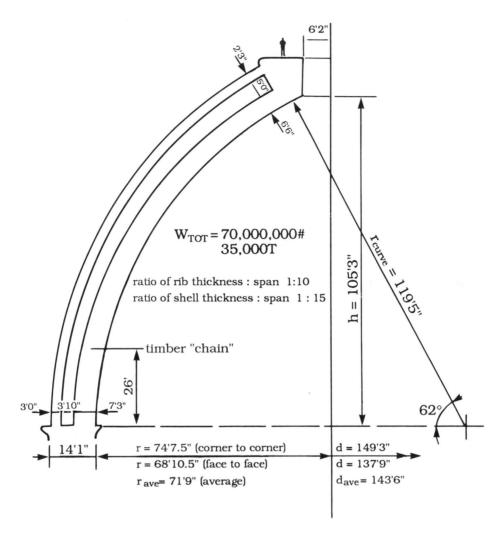

FIGURE 8-6.   Brunelleschi's Dome on the Florence Cathedral. (Drawing courtesy Professor Daniel F. MacGilvrey.)

At its spring, the dome is solid with a thickness of 4.26 meters and divides into the two shells at 3 meters height. The inner shell base is 2.2 meters and the outer 1 meter, with the inter-space 1.2 meters, increasing to five at the top (see Figure 8-7). At each angle of the octagon base there springs a rib and on each face are two secondary ribs. The two shells and 24 ribs are connected by arches at right angles to the ribs, forming a single structure. Brunelleschi added a chain to resist the outward thrust. This "chain" passed through the ribs and was composed of chestnut timbers 7 meters long, 33.6 centimeters deep and 25 centimeters wide, with square ends fastened by iron bolts. The dome was completed in 1436. Modern analyses[14] show an outward stress component at the foot of the main ribs of 816,480 kilograms, totaling 6,530,000 kilograms over the eight. The bursting tendency or hoop tension is 1,000,000 kilograms, which is resisted by the support walls. The joint of rupture is located well below the half-rise and the dome acts as a true arch, sending the forces downward rather than outward, obviating the need for the chain as designed. Still, the presence of the chain in the construction shows Brunelleschi's appreciation for the exact location of the bursting point even though he could not calculate the exact amount of force at that point.

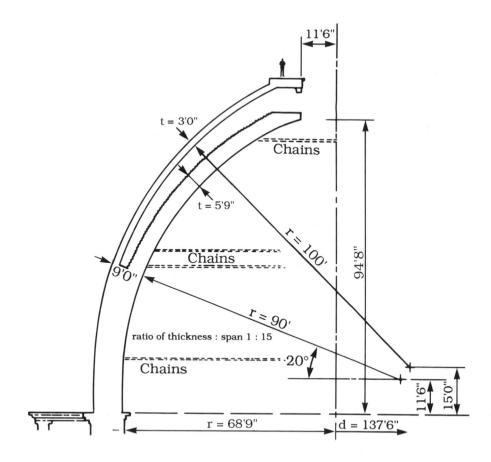

FIGURE 8-7. Michelangelo's Dome on St. Peter's Cathedral. (Drawing courtesy of Professor Daniel F. MacGilvrey.)

ST. PETER'S DOME

The second great Renaissance dome was at St. Peter's Cathedral in Rome and is by Michelangelo and Giacomo della Porta. Two men are given here as its sires as one (Michelangelo) designed it and the other (della Porta) built it.[15] Parsons, evaluating the two, considers the second work less bold in plan and smaller by comparison.[16] Pevsner, nonetheless, has the dome with the back of St. Peter's as Michelangelo's masterpiece.[17] From an architectural standpoint, if one agrees that the aesthetic is genuinely in that province alone, the St. Peter's dome is a landmark. It breaks with the Renaissance in that Michelangelo breaks with the ordered design as it appears in the High Renaissance. The dome expresses Michelangelo's personal view where balance is now counter-pointed with discord. The classical symmetry so characteristic of the Renaissance is not in his design.

Michelangelo discarded Bramante's original design as a drum similar to the Pantheon. His dome was circular rather than octagonal, lower, with a diameter of 40.8 meters compared to the Florence dome's (43.5 feet) and the Pantheon's 42.9.[18,19] The rise in the Pantheon was 21.6 meters, Santa Maria 32 meters, and St. Peter's 27.9 meters.

An example of Renaissance engineering is that of the Vatican Obelisk and the work of Domenico Fontana (1543–1607). Fontana was one of the first to be regarded as a "civil engineer" as he:

(1) Mastered difficult technical problems
(2) Designed road and hydraulic work

(3)   Worked at practical building construction
(4)   Had a knowledge of mathematics, geometry and the deliberate use of calculations.

Fontana undertook the removal of the great Egyptian obelisk which had stood by the side of St. Peter's at the site of the Circus Maximus. It was to be re-erected in the center of the great piazza in front of the new church.

A competition was held with Fontana's proposal being accepted. The project's preparation took 7 months, with large numbers of windlasses, tackles, ropes, etc., being built and assembled. The first phase was lifting the obelisk from its base and placing it horizontally (trumpets started windlasses, bells stopped them). Any violation during the critical lifting phases was dealt with stringently, even with execution. The obelisk was transported on a new, lower, earthen wall on which the obelisk moved on rollers. Some 907 men and 75 horses were used at 40 capstans to raise the obelisk and place it in its cradle over an 8 day period. The pedestal was moved over the next few months (April — September). The obelisk was raised using 800 men and 140 horses at 40 capstans. Thirteen hours and 52 minutes were used to lift it to an upright position.

Relying primarily on capstans with a 9:1 advantage and pulleys of a 2:1 advantage (of the type available to the Romans), Fontana used 40 capstans to raise 80% of the monument's mass; levers carried the balance. The equation between the muscular power of horses to that of men was roughly 10:1.[20]

Fontana calculated the weight of the obelisk by weighing in similar stone a cube of one span side length with a specific gravity of 86 Roman Pounds/cubic span or 256 kg/m^3. The weight was 963,537 35/48th Roman Pounds or 327 long tons (340 tons). He estimated that a windlass could lift 7 tons so he used 40 such windlasses (or capstans). The hoisting ropes were 7.5 centimeters in diameter. Moving the Vatican obelisk was a significant feat for Renaissance engineering but we must remember that Roman engineers removed the obelisk from Egypt, loaded it and crossed the Mediterranean, unloaded, transported, and then erected it in Rome.

Andrea Palladio (1518–1580) reintroduced the truss to modern bridge building. Short pieces were framed together to make long spans. Palladio used only one diagonal brace in a panel. This is all right if this member can carry compression and tension. In general, two braces were thought more secure. It is ironic that Palladio, an architect who, it is said, memorized Vitruvius, should be the first to describe an effective truss bridge, a form regarded by modern architects as one of the ugliest creations of the engineer (Figure 8-8).

THE ENGINEER AS SCIENTIST: TACCOLA AND DA VINCI

Marianus Jacobus, known also as Mariano Taccola (1382–1453), wrote extensively on technical concepts and machines. His two works of most interest to us are *De ingeneis* and *De machinis*.[21] He was an engineer who wrote on devices, bridges, harbors, aqueducts and construction. His works were drawn from what he was heir to at that time and what he observed. He worked as an engineer, as well as taught, being superintendent of streets in Sienna (1442). Here he applied his knowledge in civil and hydraulic construction. As an author he does not seem to have known of Hero, Archimedes or Vitruvius directly from their writing. His aqueduct designs for Sienna show characteristics of older Roman design.[22] He was a friend of Brunelleschi and he discusses the latter's inventions and patent problems.[23]

Leonardo da Vinci referred to himself simply as "engineer".[24] His greatest contribution to modern engineering was his work on the analysis of stress, mentioned previously. As we have seen from the illustration of Brunelleschi's quandary over forces in his dome, there was no knowledge regarding the strength of materials. Leonardo's work in the area of arch rupture is shown in Figure 8-9. These were not qualitative resolutions as was Brunelleschi's solution in the wooden chain. Da Vinci concludes: "An arch will not break if the chord of the outer arch does not touch the inner arch."[25]

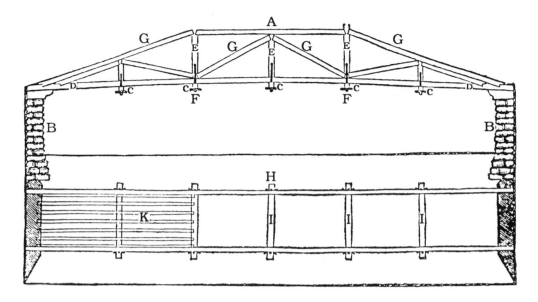

FIGURE 8-8. Palladio's Truss.

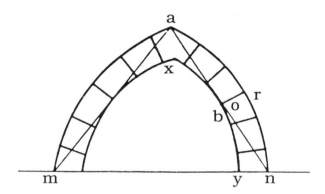

FIGURE 8-9. Leonardo's diagram of how to determine the breaking
strength of an arch. (Redrawn from Parsons, 1939.)

In other words, by reference to Figure 8-9, when "the chord *aon* of the outer arch *nra*
approaches the inner arch *xby* the arch will be weak, and it will be weak in proportion as
the inner arch passes beyond the chord."[26] This statement closely conforms to the modern
concept of the line of thrust which was empirically deducted by the Gothic builders. Leonardo's thrust diagram and solution are shown in Figure 8-10, which foreshadows the principle
of moments.

The composition and resolution of forces is directly shown as he lets the weight P be
resisted by Q and Q', which are equal weights. He concludes that the perpendicular AB is
to AC what the weight Q is to P or that if P is greater than zero it is impossible for Q to
tension AB to zero. Therefore, PP' cannot be horizontal. Herein lies the germ of analysis
of determinate structures such as trusses and suspension bridges.

Da Vinci was wrong at times. He was confused as was Frontinus by the relationship
between discharge and head. On the concept of work (force through a distance) he erred
mathematically by making space proportional to time.[27] "...The same force (that moves a
body over a given distance for a given time) will move half this mass through the same
distance in half this time." Leonardo states the relation as 2/1 rather than the correct $\sqrt{2}/1$.

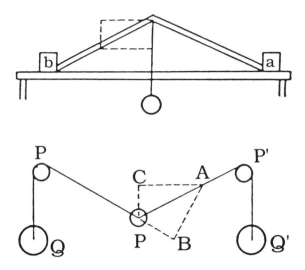

FIGURE 8-10. *Upper:* A thrust diagram by Leonardo. *Lower:* Leonardo's diagram giving the solution of the problem. (Redrawn from Parsons, 1939.)

We have little evidence of his works as a working engineer. He left no Santa Maria or St. Peter's by which we can judge his skill. He did work on the Lombard Canal System[28] and Cesare Borgia hired him for fortress design. Gille has criticized him as nonoriginal in many of his "inventions" and even of borrowing from his predecessors.[29] In this, Gille is generally correct but his conclusion that much of Leonardo's work, scientific or applied, was not universal is in error. In a way he was Taccola's successor but his breadth of inquiry is still astonishing today. His inventions may have been derivative, his engineering unbuilt and his science sometimes in error but we must appreciate where he, Taccola, Brunelleschi, Alberti and others stood — at the end of a long period of intellectual silence and the beginning of modern inquiry. Just understanding the clutter of knowledge given to them and then going beyond it was a triumph.

MATHEMATICAL, GEOMETRICAL RULES AND TOPOGRAPHICAL SURVEYING

Carlo Fontana (1634–1714) was from the Ticino of Southern Switzerland. He describes a rule for determining the shape and dimensions of a masonry dome, e.g., E (thickness of drum wall) equals 1/10th the span; the thickness of a dome at its spiring is 3/4th of its thickness (E).[30]

In surveying up to the Renaissance, the number of instruments had hardly increased since antiquity. Swiss engineers and surveyors began development and refinements that continue up to today. During the Renaissance the problem of measuring heights and distances by indirect methods was solved by elementary theorems of triangles. Loanhard Zubler (1563–1609), Zürich goldsmiths and mechanics, together with Phillip Eberhard, a Zürich mason, were the first to carry out field surveys with "playne tables" about 1600. Simon Stevin used decimal notation, speeding up calculations.

Again in surveying, the balculum or crosstaff was developed. Also used was the geometric quadrant or square (for measuring vertical angles by side ratios in degrees for leveling). The "theodolitus" was a quadrant or square mounted on a staff where a surveyor could rotate the square on a vertical angle, measure the altitude and multiply by the proper scale factor (say 120) and divide by three, obtaining the horizontal distance or corrected "slant range".

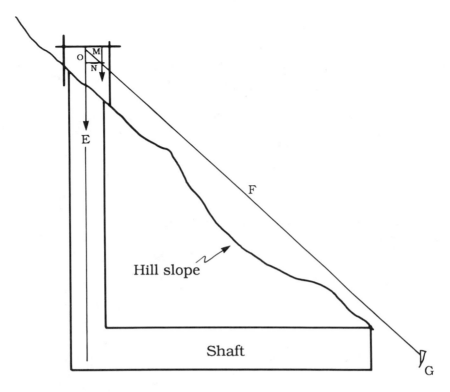

FIGURE 8-11. Agricola's illustration of Renaissance mine surveying; MNO is proportional to triangle formed by altitude of the shaft E, the length of the shaft EG, and the cord F. (Redrawn from Dibner, 1958.)

Table 8-1

1 Miglio = 1000 passi =	1628.97	yards	or	1489.4788 m
1 Catena (land surveyors) =	42.149	feet	or	12.8468 m
1 Catena (engineers) =	36.651	feet	or	11.1711 m
1 Passo =	4.887	feet	or	1.4895 m
1 Braccio, commercial =	33.387	inches	or	0.8482 m
1 Piede =	11.729	inches	or	0.2979 m
1 Palmo (engineers) =	8.796	inches	or	0.2234 m

Edmund Gunter (1581–1626), an Englishman, devised the "chain" of 10 square chains/acre. He also invented a forerunner of the slide rule in a "graphical logarithmic scale". He developed tables for series, tangents and logarithms.

Georgius Agricola (Georg Bauer) (1494–1555) wrote *De Re Metallica* ("of Things Metallic"). One of the greatest early engineering books, it is a pioneer work in chemical engineering and mining engineering — 10 editions in three languages. It was the outstanding work until 1738 on furnaces and metallurgy, with discussions on mining methods little improved until the introduction of steam, drills, and dynamite. Agricola pioneered in finding of ores, the theory of formation of mineral veins; mining law; mine surveying; tools, machines, methods; pumps, hoists, water power; ore preparation and smelting; and manufacture of salt, soda, alum, vitriol, sulphur, bitumen and glass.

The English translation was by Herbert Hoover, a famed mining engineer in his own right, before he became an American president. His discussion of mine surveying clearly shows that the Renaissance knowledge had advanced beyond that of antiquity[31] (see Figure 8-11). Typical units used by surveyors of the 15th and 16th centuries are shown in Table 8-1.

NOTES

1. **Burke, J.** *The Day the Universe Changed.* Little, Brown (1985), p. 55.
2. *Quest for the Past.* Dorling Kindersley, Ltd., Eds. Reader's Digest Association, Montreal (1984), pp. 277–278.
3. Ibid., p. 292. The Donation of Constantine War.
4. **Finch, J. K.** *The Story of Engineering.* Doubleday-Anchor (1960), pp. 115–119.
5. **Parsons, W. B.** *Engineers and Engineering in the Renaissance.* MIT Press (1939), p. 372.
6. *Trattato dei pondi, leve e triari.* (Codice Laurenziano, No. 361, Serie Ashburnham), now in the Laurentian Library at Florence.
7. **Alberti, L. B.** *De re aedificatoria.* N. Lavrentic (1485). Copy in Dibner Collection, National Museum of History, Washington, D.C.
8. **Gille, B.** *Engineers of the Renaissance,* MIT Press (1966), p. 92.
9. **Parsons,** op. cit., p. 587.
10. Ibid.
11. **Pevsner, N.** *An Outline of European Architecture.* Pelican (1968), p. 200.
12. **Gille,** op. cit., p. 80.
13. **Parsons,** op. cit., p. 590.
14. **Parsons,** op. cit., pp. 594–599.
15. **Pevsner,** op. cit., p. 230.
16. **Parsons,** op. cit., p. 611.
17. **Pevsner,** op. cit., p. 229.
18. **Finch,** op. cit., p. 112.
19. **Parsons,** op. cit., p. 590.
20. **Garrison, E. G.** Engineering puzzle solved. *Civil Engineering,* Vol. 54, No. 6, ASCE (1984), p. 30.
21. **Prager, F. D. and Scaglia, G.,** *Mariano Taccola and His Book DE INGENEIS,* MIT Press (1972).
22. Ibid., p. 55.
23. **Prager, F. D.** A manuscript of Taccola quoting Brunelleschi, on problems of inventors and builders. *Proceedings, American Philosophical Society,* CXII (1968), p. 131.
24. **Parsons,** op. cit., p. 67.
25. *Leonardo da Vinci's Notebooks English Arranged and Rendered Into English by Edward McAudy.* New York (1923).
26. Ibid., Ms.E. 60v.
27. **Hart, I.,** *The Mechanical Investigations of Leonardo da Vinci.* London (1925).
28. **Carvill, J.** *Famous Names in Engineering.* Butterworths (1981), p. 45.
29. **Gille,** op. cit., p. 143.
30. **Bélidor.** *Science des Ingénieurs.* Paris (1729); in Straub (1964), p. 92.
31. **Parsons,** op. cit., p. 19.

9 17TH AND 18TH CENTURIES

THE FRENCH AND ENGLISH

For all the Renaissance revival in Italy, its engineering provided little to the practical art of engineering after the 16th century. France became the center of engineering from the 17th century. In science, France must share the stage with the English — notably due to Sir Isaac Newton.

Isaac Newton (1642–1727) was the greatest scientist of the 17th and 18th centuries. Apart from the theory of gravitation and optical theories he was co-inventor with Leibnitz[1] of the calculus. His work in classical mechanics produced:

1. The generalization of the concept of "Force"
2. The formulation of the concept of "Mass" (First Law)
3. The Principle of Effect and Counter-Effect (Third Law)

Leonhard Euler (1707–1783), a Swiss mathematician, furthered the Basle school's (he was one of Leibnitz's students) work in bending. His work on buckling of struts resulted in the formula

$$P = \frac{\pi^2 EI}{c_i^2}$$

Here, as in other work such as that on catenary curves, the primary aim was not to solve buckling problems or determine elastic curves and associated centers of gravity, but to apply the newly invented calculus. Euler proved that the short columns fail by compression; long ones by bending. Euler published the formula in 1757.[2]

$$P = \frac{Kd^4}{l^2}$$

He was the first to formulate laws governing the flow of fluids and to explain the importance of pressure to flow, a relationship correctly noted by only one worker in antiquity, Hero. In other bending studies Robert Hooke (1635–1703), a British physicist, determined that elasticity was critical to a complete evaluation of displacement, i.e., force with which a spring attempts to regain its natural position is proportional to the distance which it has been displaced. Stress = F/A (force per area); Strain = x/h or $\Delta l/l$ (where l = length); Y = stress/strain, which leads to Young's Modulus or

$$\text{Force Constant} = \frac{YA}{L_o} \Delta L$$

Hooke analyzed the forces acting on the arch and formulated a limit of elasticity in materials. The "universal joint" for the transmission of power from one shaft to another connected to it at an angle was another of Hooke's contributions to engineering.[3]

Charles Auguste Coulomb (1736–1806), the great French engineer and physicist for whom the Coulomb unit was named, wrote the famous Mémoire on structural analyses: "Essai sur une application de règles de maximis et minimis à quelques problèms de statique relatifs à l'architecture''. In this he formalizes the universal treatment of the cantilever beam of rectangular cross-section. His treatment takes into consideration the shearing strength as well as compressive and tensile strength, and admits, in principle, a relationship between stress and strain. The special case of the perfectly elastic body is:

$$M = \delta \frac{bh^2}{6}$$

Coulomb in his paper on statics summarizes three basic laws of statics and equilibrium:[4]

1.   The sum of the tensions must balance the sum of the compressions.
2.   The sum of the vertical components of the internal forces must equal the load applied.
3.   The sum of the moments of internal forces must balance the bending moment produced by the applied load.

By 1779 Coulomb suggested the use of compressed air in cassions. Application of this technique by later engineers led to construction of bridges across great rivers and bays, together with tunnels in areas never before attempted.

Pierre Fermat (1601–1665) and René Descartes (1590–1650) independently invented analytical geometry. Newton in *Principia* stated his now famous laws: *(Philosophie naturalis principia mathematica)*[5]

1.   Every body continues at rest, or in uniform motion, unless compelled to change that state by forces impressed upon it.
2.   The change of motion is proportional to forces impressed upon it.
3.   To every action there is always an opposite and equal reaction.

Edmé Mariotte (1620–1684) had correctly defined the ''axe d'equilibré'' or neutral axis of a beam, which preceded Coulomb's work.

REPAIR OF ST. PETER'S DOME — THEORY VS. PRACTICE

By the middle of the 18th century, the principle of virtual displacements was known.[6] The properties of many important building materials were tabulated. Further, the statical behavior of buildings was being examined by rationalist scientists. This era was marked by the first attempt to theoretically survey the stability of structures. St. Peter's was originally the design of Donato Bramante (1444–1514), which was a monumental plan of a church based on the Greek cross of equal length arms surmounted by a great dome. His dome was circular, resting on four great columns. He envisaged a semi-circular dome like that of the Pantheon. At his death Bramante's nephew, the great painter Raphael Sanzio da Urbino, was placed in charge. Neither architect nor engineer he held the post for 5 years. Guiliano San Gallo succeeded him and continued Raphael's strengthening of Bramante's columns, which were underdesigned.[7] Michelangelo, who next took up the commission, further strengthened the columns, as can be seen in Figure 9-1. The dome was redesigned by San Gallo to that of a rotated segmental arch which omitted step-rings such as used in the semi-circular dome, and substituted longitudinal ribs. He retained the single-shell design. As we have seen, Michelangelo kept San Gallo's higher dome but his was a double-shell design which his successor, Giacomo della Porta built in 1588. The choice of a circular segmental design with a large stone lantern produced a construction which allowed the thrust line to exit higher on the dome, just as in the Romanesque designs. Less of a Gothic dome than Brunelleschi's, the dome developed cracks almost immediately. In 1742–1743, Pope Benedict XIV asked for a structural analysis of St. Peter's Dome, in order to determine the

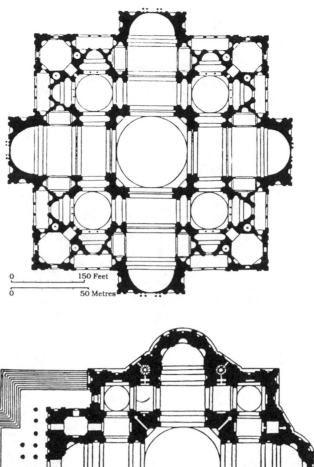

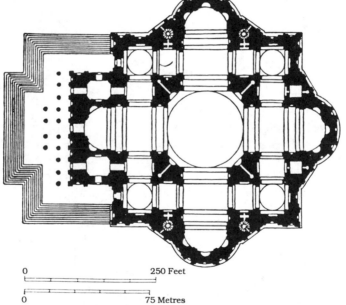

FIGURE 9-1. *Upper:* Bramante's original plan for St. Peter's, Rome, 1506.
Lower: Plan for the completion of St. Peter's, Rome, by Michelangelo, 1546.
(Redrawn from Pevsner, 1968.)

cause of the cracks and damage. Three authors — Le Seur, Jacquier, and Boscowich —
evaluated the design of the building, and determined the impost ring of the dome as the
problem. The interesting part then began, with an attempt to calculate the horizontal thrust,
and to prove the two iron tie rings, or bands, built into the dome, were no longer able to
withstand that thrust.

Using the principle of virtual displacements after Jordanus, Descartes, and Bernoulli,
they regarded the fractures as a moving hinge joint (see Figure 9-2), with the ratio of relative

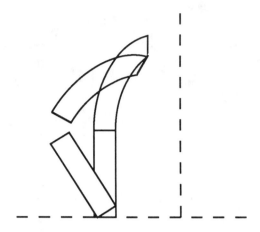

FIGURE 9-2.   Fracture of a moving joint.

displacements obtained geometrically from the graph. The three workers assumed, incorrectly, that all components of the system are geometrically rigid. They had no idea of elasticity. They correctly assumed the iron rings would expand, correlating the fractures with the elongation of the iron. The resistance of the tie rings (which grows with elongation) was erroneously treated as a constant force. (They had no knowledge of Hooke's pioneering work.)

The authors assume a total thrust spread over the circumference where

$$H = \Sigma G \frac{v}{h}$$

G = the weights; V = the sag at center of gravity; and h = the horizontal displacement of the impost ring. H is opposed by the resistance, W, composed of the tilting resistance of the drum wall, buttresses, and tie rings. The longitudinal tensile force, Z = pr or $H/2\pi$, where the radial thrust equals p, whereby the authors wrongly calculated a deficiency in horizontal resistance at the impost height of 3,237,356 Roman Pounds (~1100 tons) which was used to recommend a remedy by use of additional tie rings. A safety factor of two was taken into consideration.

Criticism of the report and analysis was immediate and vitriolic — "if it was possible to design and build St. Peter's dome without mathematics,...it will also be possible to restore it without the aid of mathematics... Michelangelo knew no mathematics and was yet able to build the dome... Mathematics is a most respectable science, but in this case it has been abased...". The most correct objection was to the magnitude of the deficiency in horizontal resistance (1.25 million kilograms). The authors failed to stress adequately that the stated value was a maximum value.

Less important but perhaps somewhat valid is the recognition of differential settling, poor masonry, earthquakes, etc., as sources of fracture as well, by Giovanni Poleni of Venice. Poleni agreed with the three scientists as to the corrective measures of more tie rings. Five rings were used in the repair. The importance of the study is its break with the past in its use of a scientific analysis of the structural problem.

Sir Christopher Wren avoided the problems of St. Peter's in his unique dome design for St. Paul's cathedral which replaced Old St. Paul's, destroyed in the Great Fire of London, 1666. Wren (1632–1723) was a distinguished mathematician by age 28, before he was an architect, having studied mechanics, hydraulics, and astronomy.[8] Unlike Brunelleschi, Michelangelo and the other builders of the late and Post-Renaissance, Wren *was* in touch with

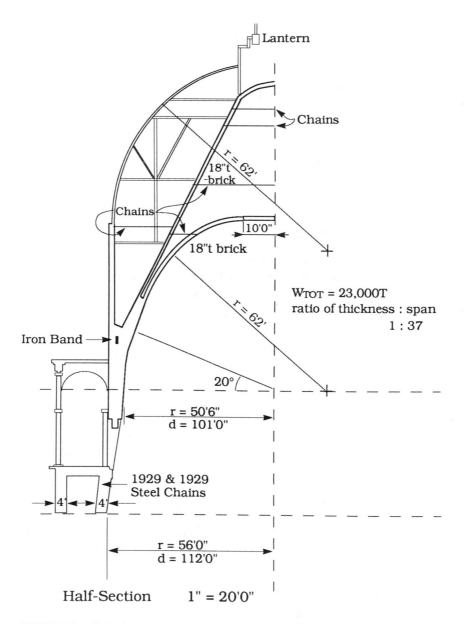

FIGURE 9-3.   Sir Christopher Wren's Dome on St. Paul's Cathedral. (Courtesy Professor Daniel F. MacGilvrey.)

the works of men like Alberti, Stevin, Galileo, Leibnitz, Pascal and certainly Newton, as well as the major architectural treatises of the Italian Renaissance.[9,10]

St. Paul's, built between 1675 and 1710, performs the role of St. Peter's in the Anglican church. Both are monumental structures topped by great domes. The chief divergence between the two lies in their structural difference, which traces to the one harking backward to a classical, empirical tradition of building design vs. an experimental approach based on theoretical concepts about structures newly proposed in the 17th century. Wren's dome (Figure 9-3) is a three-shell dome. The one most obvious to the viewer is the lead-sheathed, wooden frame outer dome, which forms an integral part of London's skyline. The inner dome was a brick construction that is open, allowing a view of the lantern. The genius of Wren's dome lies in the middle dome, which is a brick cone supporting the pressure of the

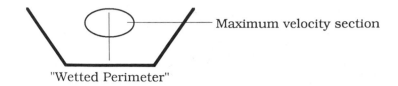

FIGURE 9-4.   Maximum current velocity in an open channel.

850 ton lantern, as well as the lighter outer dome. The device of brick cone mimics the Gothic arch in its ability to transfer the line of thrust more directly to the supporting ring and walls.[11]

FRENCH ENGINEERING SCHOOLS, CANALS, BRIDGES, AND ROADS
Schools

Sébastien la Prestre de Vauban (1633–1707), the great French military engineer, suggested a Corps of Engineers be organized (by the Army) to build roads, bridges, and fortifications. He personally built the harbor at Dunkirk into a fortified port and harbor (1671–1683). Vauban:

- brought polygonal and star-shaped fortifications to perfection
- was a soldier/military engineer
- developed canals as an assist to the defense of country
- completed the Canal du Languedoc.

During his life the term "Ingénieur" was used for the first time.

Jean Baptiste Colbert (1619–1683), who was minister to Louis XIV, founded Corps de Genie in 1675. This concept was first suggested by Sébastien la Prestre de Vauban. The Corps de Ponts et Chaussées was created in 1716. Shortly thereafter, the École des Ponts et Chaussées was founded in 1747 and reorganized by Jean Perronet in 1760. *Science des Ingénieurs* by Bernard de Belidor was published from 1729 until 1830. Napoleon authorized founding the École Centrale des Travaux Publics in 1794 and the École Polytechnique in 1795 for advanced design studies taken at École des Ponts et Chaussées. L'École Polytechnique has been the model for engineering schools since that time. Important examples include:

- Eidgenössisches Polytechnium, Zürich — 1855
- Polytechnique Schools, Delft — 1864
- Massachusetts Institute of Technology (MIT) — 1865
- University College, London — 1828.

Canals and Hydraulics

Evangelista Toricelli (1608–1647) determined that the velocity of flow varies with the depth, i.e., $\sqrt{d}$ or head of water on the opening. ($\gamma = \sqrt{2gh}$ or $\gamma = \sqrt{2gd}$ after Bernoulli). Applied to stream flow velocity, this posited that the velocity of water in a flowing stream increased with depth, which was wrong. Marotte determined that the flow was only $^7/_{10}$ the theoretical value. DuBuat and Chézy reversed this theory, correctly observing that maximum flow velocity is the central channel at surface to mid-depth (Figure 9-4).

David Bernoulli (1700–1782) derived the concept of the relationship of flow velocity and cross-section, i.e., $Q = V \times A$ (cf. Hero) to its form:

$$V = \sqrt{2gh}$$

Which is the special case of Bernoulli's general equation:

$$p + pgy + \frac{1}{2} pv^2 = C$$

This knowledge was extended to open channels, ditches, canals, and streams by Antoine de Chézy (1718–1798) with the development of "the Chézy Formula":

$$V = C\sqrt{DS}$$

where V is the average flow velocity; D is the mean depth of the channel; and S is its slope. D was later replaced by the more widely applicable dimension, R, or "hydraulic radius". C is a constant relating R with S and friction. Henri Pitot (1695–1771) developed a device (the "Pitot tube") for measuring pressure and velocity of a current. Pitot was typical of French engineers in his combination of theoretical training and application.

French engineers built canals to lift traffic from lower to higher systems, i.e., barge and ship canals. Further, they used locks to deepen ports and entrance channels by opening the locks at low tide and releasing the inner port water at high speed, e.g., "hydraulic dredging". The French used caissons and cofferdams in marine and riverine constructions. French dams were built to feed the canals, such as the earthen dam at Saint-Ferreol at 32 meters in height with three core dams of masonry, while the earliest modern era masonry dam was built in Spain (1580) to 41.1 meters at Alicante, to be used for irrigation.

Canal du Languedoc

This canal was part of a system of strategic barge canals that Vauban saw in much the way military planners view the great autoroutes such as the German "auto bahn" or U.S. "interstate". Colbert, again showing his appreciation of engineering, convinced Louis XIV to construct the Canal du Midi or Languedoc.[12] Its construction followed two earlier canals — the Braire Canal (1605–1642), only 55 kilometers long, which joined the Loire and Seine Rivers, and the Canal de Orléans, a 74 kilometer canal dug for barges moving up the Loire to Orléans. The Braire Canal was the first canal in France to have the Italian-inspired locks (four) and the first to cross a divide (81 meters).[13]

The Languedoc or *Canal des deux mers* was completed during the reign of Louis XIV and called by Voltaire the most important monument of that great French king.[14] The canal (called Canal du Midi today) was built by a retired tax collector turned civil engineer, Pierre-Paul Riquet (1604–1680). Riquet was given his commission in October, 1666, after completing a 42 kilometer feeder canal to the summit of the main canal. The route of the main canal ran from Toulouse on the Garonne River to the Mediterranean at Sète.[15] The canal is 232 kilometers long, rising to an altitude of 189 meters by means of 74 locks (Figure 9-5, upper) and descends via 26.[16] The canal bottom width averages 10 meters, with a depth of 2 meters, allowing passage of 200-ton barges. The canal was carried over intervening streams by three elegant viaducts (Figure 9-5, lower). The side-slope is 2.5 to 1, yielding a surface width of 20 meters. The lock lengths were 35 meters with 6.5 meter gates. The canal features a 55 meter tunnel near Bézier and 32 meter high dam with masonry core at St. Ferréol that stored 7 million cubic meters of feed water for the summer months. The canal project broke Riquet's health and left him penniless; he died in October, 1680, 7 months before the opening of the canal. The Languedoc Canal was the greatest hydraulic engineering achievement of the 17th century, being monument to both Sun King and his engineer. In later centuries the canal was extended to Bordeaux and lengthened at the Mediterranean, reaching across southern France and performing for that nation what a later 20th century canal would do for the Americas — connect two oceans.

A less notable hydraulic engineering achievement during Louis XIV's long reign, was by the Flemish engineer Rennequin Sualem (1645–1708), who built the Versailles water system to raise water 169 meters by 225 pumps, powered by 14 undershot wheels. Only

Plan de deux Sas ac-
colles dans le gout de
ceux de Foucerane pro
che Bezieres au Canal
de Languedoc

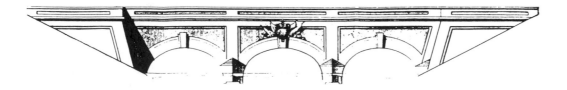

FIGURE 9-5.   *Upper:* Locks of the Languedoc Canal. *Lower:* Trebe River viaduct along canal. (From Sandström, 1970.)

5% efficient, it utilized cast iron pipes for the first time for transmission lines. Jean Lintlaer, another Flemish engineer, built 5 meter water wheels under the Pont Neuf in Paris in 1608 to pump water from the Seine to the palaces of the Louvre and Tuileries. These were called the "Samaritaine" (Figure 9-6) and lasted up to Napoleon's reign (1813).

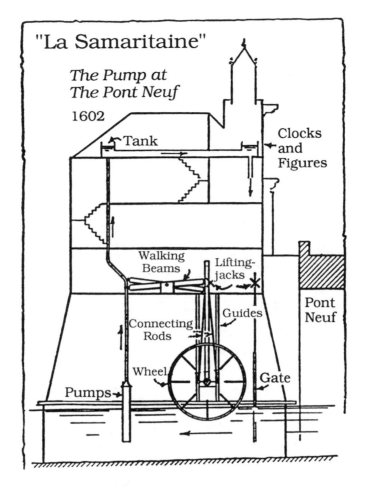

FIGURE 9-6. "La Samaritaine". (Redrawn from Finch, 1960.)

French Bridges

Max Bill, writing on bridge design, credits Jean-Rodolphe Perronet (1708–1794) with the creation of "the highest and most typical expression of stone bridges".[17]

Characteristics of Perronet's bridges include:

1. Corne de vache ("cow's horn") pier design which increased flow of water through arches by the oblique flattening of the openings (Figure 9-7)
2. Segmental, longer-span flat arches on the narrower piers (e.g., Ponte Vecchio) used by Perronet to reduce obstruction to flood flow inherent in stone arch design

Perronet used hard stone to handle the high compression in his flat arches. On the Pont de la Concorde in Paris the piers become so slender as to allow 65% of the waterway open to flow (Figure 9-8). The arches rise only 4 meters over a length of 31 meters. Chézy completed Perronet's great design in 1791.

Perronet served as chief of both the Corps d'Ingénieurs and the head of the École des Ponts et Chaussées. He worked on the Canal de Bourgogne and wrote extensively on pile foundations and driving resistance.[18]

Louis Marie Henri Navier (1785–1836) created building statics or structural analysis. As a great teacher he developed the theory of flexure, the theory of buckling, and solutions of statically indeterminate problems. In 1824 he formulated the deflection and strength of a beam in correct terms where stress and strain are proportional. As great an engineering scientist as Navier was, his greatest failure was in bridge design.

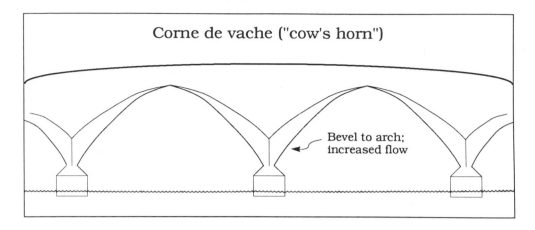

FIGURE 9-7.   Elevation of Perronet's bridge pier design.

Pont de la Concorde, Paris   Waterway below floor level 65%   French 1787 - 1791

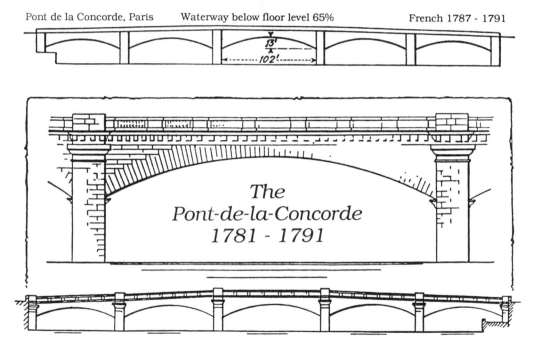

FIGURE 9-8.   The Pont de la Concorde, 1787–1791. (Redrawn from Finch, 1960.)

Navier built several bridges over the Seine.[19] His failure was a suspension bridge over that river in Paris. A detail of the anchors is shown in Figure 9-9. The structure, almost complete, had to be dismantled and was not used. His contemporaries blamed poor subsoil, difficult water drainage, and jealousies within the City Council rather than Navier.[20] Still, the great engineer writes: "To undertake a great work and especially a work of a novel type, means carrying out an experiment. It means taking up a struggle with the forces of nature without the assurance of emerging as the victor...".[21]

Outside of France one of the largest wooden bridges was built by the Swiss Johann Ulrich Grubenmann (1709–1783). It crossed the Rhine at Schaffausen (Switzerland) with a 119 meter span at one point. Authorities did not trust his design. He changed his design, placing it on a foundation in the middle of the river. After construction he took away the

FIGURE 9-9.   Navier's Seine bridge. (From Straub, 1964.)

FIGURE 9-10.   Grubenmann wooden bridge design, ca. 1764. (From Billington, M. and Abel, Eds., *The Maillart Papers,* Princeton, 1973.)

wooden blocks between the foundation and the bridge to show the strength of the now 106 meter span. It was destroyed by French soldiers in 1799. A similar design by Grubenmann is shown in Figure 9-10.

French and British Roads

Before the 18th century, France had about 50,000 kilometers of roads maintained by corveé labor — men ''drafted'' for up to 50 days from farms to work on roads. In Britain the situation was worse.[22] The situation was such that the Royal Navy had difficulty obtaining masts for its vessels due to the ''exceeding badness of the roads''. Still, in France there lay a tradition of public maintenance of roads that can be traced to Charlemagne.[23] This monarch had planned to repair or reconstruct the old Roman system of roads. After his death these efforts foundered for centuries, up to the Renaissance.

Interest in road construction grew again in 1464 when Louis XI created a Royal Courier Service. This act laid the foundation for expanded communication. By the 16th century a regular tariff was imposed and tolls were used for highway repair, following an ordinance by Louis XII (1508). Two types of roads were described: royal and private.[24] Royal roads varied from 30 pieds (feet) to 60 pieds, depending on terrain. Typically, only a 15 pied was "paved". Secondary roads followed a classification system much like that seen with the Romans. A "path" was 4 pieds wide; the "cart" road was 8 pieds wide; the "way" was 16 pieds wide; and the "road" was wide enough for two carts to pass. Paving, such as used in Paris, was of cubical blocks 7 to 8 pouces a side. For maintenance the "corvée" was adopted, by which men were called for up to 50 days of road labor.[25] This was not the best method for road maintenance and was abolished in 1776 by Louis XVI.

Pierre Marie Jérôme Trésaguet (1716–96) presented a Memoir to the Corps des Ponts et Chaussées, in 1764. Trésaguet realized, as had Guido Baldo Toglietta around 1585, that there were two elements to be considered in road design surface: material and foundation.[26] Both these men recommended the use of gravel on crowned, well-drained road bed. It was Trésaguet who actually applied these principles. In comparison to the Roman road, his road was less deep (ca. 25 centimeters) and made of stone carefully graded from a foundation of stones set on edge, through a middle layer of wedged stones, which grades into a final, upper layer 7.5 centimeters thick of hard gravel (Figure 9-11). As with Toglietta, he counseled systematic maintenance and repair. Trésaguet's road could be built for half the cost of previous ones and had a wearing surface of 10 years or more depending upon maintenance.

Thomas Telford (1757–1834) and John Loudon MacAdam (1756–1836) were to do for Britain what Trésaguet had done for France. Telford modified Trésaguet's road to make it last longer, but in the process increased its cost.[27] Telford approached road construction like the mason he was before his architecture and engineering career. Much like Roman roads, his lower courses were laid to carry the weight of traffic no matter the quality of the soil. As such, his lower course was flat (Figure 9-11), with the other two layers grading into a rammed gravel surface. The Telford road was 10 meters in right-of-way, with the road itself about 6 meters wide. Telford designed embankment slopes at 1.5:1 while fill sections were left at elevations that allowed for settling. Telford's finest road was the Holyhead Road from London to Holyhead, Wales, a distance of nearly 500 kilometers (Figure 9-12). It was on this road where it crossed Menai Strait that Telford built his famous suspension bridge.[28]

MacAdam has become immortalized in that his name became the synonym for the modern road. MacAdam's roads differed from those of Trésaguet and Telford in his insistence on a thoroughly drained subsoil. By maintaining a dry subsoil the foundations could be made shallower, reducing costs. MacAdam transferred the load bearing to the subsoil rather than the road itself as Telford, in particular, had it. His roads were rarely over 12 centimeters thick, with rammed gravel that rain could not penetrate. He further crowned his road so rain would run off, obviating any standing. MacAdam's road design was used around the world and is the standard for well-engineered secondary roads even today.

OTHER BRITISH ENGINEERS

John Smeaton (1724–92), who first styled himself as a civil engineer, built Eddystone lighthouse, designed earthen embankment dams, invented a diving bell, and worked on steam pumps, improving Thomas Newcomen's engine.[29] Smeaton was well educated and apprenticed as a scientific instrument maker much as James Watt did (c.f. Chapter 10). He began his distinguished career with the Eddystone lighthouse, built on a reef 22 kilometers off Plymouth. The previous design, a pitch-covered shaft, burned in 1755 after a 50 year use-life.[30]

Smeaton's design was that of a masonry "tree", circular and of decreasing diameter, surmounted by the light.[31] The foundation stones weighed 2.5 tons, although he did rely on

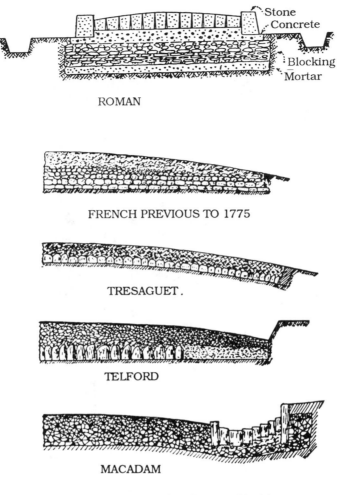

FIGURE 9-11. Road Sections, Roman to MacAdam.

their megalithic character. He dovetailed the stones to add a unitary strength to the overall structure. To add additional strength, he used a pozzalana concrete capable of setting underwater. He based his choice on empirical testing of various mixtures of lime, clays, lias and pozzalana.

Smeaton founded a line of British engineers who would overtake and excel their French counterparts. These men include: William Jessup, who built London's great docks, Thomas Telford, and John Rennie (1761–1821). James Brindley (1716–1772) was a notable exception to this lineage and was England's foremost canal engineer,[32] although he and Smeaton cooperated on several canal projects. Canals in 18th century Britain were not the strategic devices Vauban envisaged for France but became the transport routes for the Industrial Revolution. Coal required the barge canals that would carry this fuel to the growing industries of the late 18th century.

Coal had replaced wood as the principal fuel by the late 16th century in Britain.[33] Coal changed the manufacture of items from glass to iron due to the gases given off in its combustion. Reverbertory furnaces were developed, as was cementation of wrought iron to steel. Smelting and metal working benefited most from the adoption of coal. The mining of coal led to the development of steam power and the birth of modern mechanical engineering, as we shall see in the next chapter. As for civil engineering, the canal era began

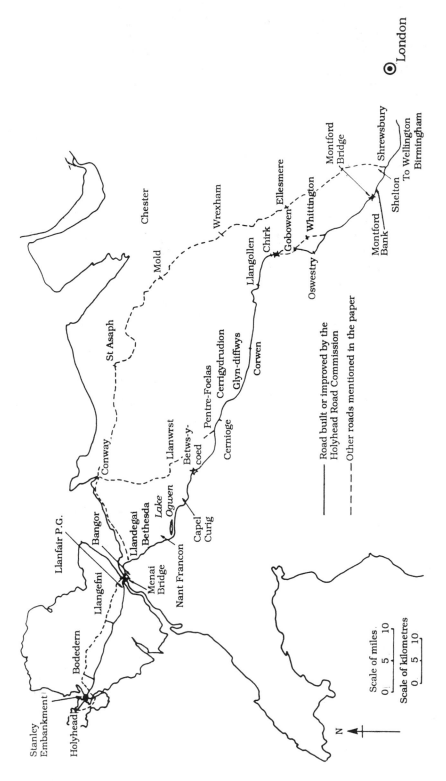

FIGURE 9-12.   Map of route of Holyhead Road. (Redrawn from Pugsley, 1980.)

in England later than in France and Europe was to produce the generation of engineers that forever linked industry and engineering.

Brindley was self taught, as were many of the English engineers. His first project was to unite the Trent and Mersey rivers. Next he was hired by the Duke of Bridgewater, Francis Eggerton, to build what was to be called the ''Worsley Canal'', from his estate at Worsley to Manchester, a distance of 17 kilometers.[34] Along its route he built the 181 meter Barton Aqueduct which allowed barges to cross the Irwell river 12 meters below.[35] The canal was extended to the Mersey in 1762, an additional 40 kilometers. He then began the Grand Trunk Canal from the Mersey to the Severn via the Trent river, which was certainly in the class of the Languedoc, being 225 kilometers in length, with 75 locks, several aqueducts, and five tunnels. Brindley preferred tunnels over deep cuts at water sheds. On the Grand Trunk his longest tunnel was 2.6 kilometers at Hardcastle. Its dimensions were but 3.6 by 2.7 meters, which led to a larger tunnel by Telford in 1827. It is ironic that Telford's great successes in canal design, including the first major cast iron aqueduct, Longdon, were followed by his greatest failures also in canals — the Caledonian Canal and the Göta.[36-38] The Caledonian, designed to eliminate dangerous navigation north of Scotland, never met its aim as the depth was reduced from 7 meters to less than 4. The Göta, a 193 kilometer canal, with 58 locks connected the Baltic with the North Sea. Completed in 1832, the canal never became a major route and did not stimulate industrial development. Telford's design was built at over six times the original estimate. The canal was less an engineering failure than an economic goof.

ADVANCES IN SURVEYING AND SURVEYING INSTRUMENTS

Simon Stevin wrote his treatise on fractions in 1586, followed shortly by John Napier (1550–1617), who invented logs in 1594 and published logarithms in 1614. Napier's student Henry Briggs (1561–1636) set them to base 10. In France Pierre Vernier (1580–1637) invented the vernier (1630) for angle measurement, which was used by Abbé Jean Picard (1620–1682) to measure arc by triangulation, giving birth to geodesic survey. Willebrord Snell von Roijen (1581–1626) was the first to use triangulation and the first to distinguish geodesic surveying from plane surveying.

By 1702 the spirit level or ''dumpy level'', a telescopic sight with level bubble, was invented by Alain Manesson Mallet (Figure 9-13). By 1723 we have descriptions in France of rules, compasses, dividers, protractors, rods, chains, pins, and tripods used for survey. The compass became a standard land surveying instrument, together with Gunter's Chain (1620) of 100 links, 20 meters long, until the 19th century.

FIGURE 9-13.   The Spirit of French Level of M. Mallet, 1702. (Redrawn from Finch, 1960.)

NOTES

1. Leibnitz (1646–1716) was more concerned with continuity and the concept of function in his calculus.
2. Foreshadowed by Musshenbroek (1691–1761), Leyden: "building resistance of struts (slender) decreases with inverse ratio of square of their length..."

$$P = \frac{Kd^2}{1^2}$$

3. **Kirby, et al.** *Engineering in History,* McGraw Hill (1956), p. 129.
4. **Coulomb,** "Essais sur une application des regles de maximis et minimus a quelques problèmes de statique relatifs à l'architecture." Academie des Ponts et Chaussées (1773). Coulomb's basic laws owe much to the work of Pierre Varignon (1654–1722); the French mathematician knew the "Parallelogram of Velocities" could be applied to forces, a Vector Concept of Force, "The moment of the resultant of two forces, related to any given point, is equal to the algebraic sum of the moments of the two component forces." This is expressed mathematically as follows:

$$R = \sqrt{R_x^2 + R_y^2}$$

where $R_x = \Sigma F_x$ and $R_y = \Sigma F_y$, and where components in F are $F_1 = F_{1_x} = F_1 \cos \theta$, $F_{1_y} = F_1 \sin \theta$, $F_2 = F_{2_x} = F_2 \cos \theta$, $F_{2_y} = F_2 \sin \theta$ etc., or generally in $C = A + B$

$$C^2 = A^2 + B^2 - 2AB \cos \theta.$$

5. **Newton, I.** *Philosophie Naturalis Prinicipal Mathematica,* John Streater (1967) in: Dibner Collection, National Museum of History.
6. Stated by Johann Bernoulli (1667–1748) in *Nouvelle Mécanique* ...''Concevez plusiers forces différentes qui agissent suivant différentes tendances ou direction pour tenir en équilibre un point, une ligne, une surface, on un corps; concevez aussi que petit movement...chacune de ces forces avancera ou reculera dans une direction...ce que j'appelle vîtesse virtual...En tout équilibre de forces quelconques...la somme des Energies négatives prises affirmativement.''
7. **Parsons, W. B.** *Engineers and Engineering in the Renaissance,* MIT Press (1939), p. 613.
8. **Straub, H.** *A History of Civil Engineering,* MIT Press (1964), p. 71.
9. **Sekler, E. F.** *Wren and His Place in European Architecture.* MacMillan (1956), p. 120.
10. **Pevsner, N.** *An Outline of European Architecture,* Pelican (1968), p. 319.
11. **Mark, R.** Structural experimentation in Gothic architecture, *Am. Sci.* **66**(3): 550 (1978).
12. **Kirby et al.,** op. cit., p. 141.
13. Ibid.
14. **Sandström, G. E.** *Man the Builder,* McGraw-Hill (1970), pp. 207–208.
15. **Straub,** op. cit., p. 133.
16. **Bélidor, B. F.** *Architecture Hydraulique,* Vol. IV. Paris (1750–1782).
17. **Bill, M.** Maillart and the artistic expression of concrete construction, *The Maillart Papers* from The Second National Conference on Civil Engineering: History, Heritage and the Humanities. October 1972. Princeton (1973), p. 6.
18. **Straub,** op. cit., p. 126.
19. Ibid., pp. 156–157.
20. Ibid.
21. **Stüssi, F.,** *Schweizerische Bauzeitung,* Vol. 116 (1940), p. 204.
22. **Defoe, D.,** *A Tour thro' the Whole Island of Great Britain,* 2 Vol. Peter Davis (1927), pp. 518, 530.
23. **Parsons,** op. cit., p. 291.
24. Ibid., p. 298.
25. **Finch, J. K.,** *The Story of Engineering,* Doubleday-Anchor (1960), p. 175.
26. **Parsons,** op. cit., p. 292.
27. **Sandström,** op. cit., p. 200.
28. **Trinder, B.** The Holyhead Road: an engineering project in this social context. In: *Thomas Telford: Engineer,* Penfold, A., Ed. Telford (1980), p. 43.
29. **Finch,** op. cit., p. 193.
30. **Sandström,** op. cit., p. 134.
31. **Smeaton, J.** *A Narrative of the Building and a Description of the Construction of the Eddystone Lighthouse with Stone.* London (1791).
32. **Straub,** op. cit., p. 134.
33. **Nef, J. U.,** An early energy crisis and its consequences. *Civilization.* W. H. Freeman (1979), p. 130.
34. **Sandström,** op. cit., p. 202.
35. **Finch,** op. cit., p. 191.
36. **Clayton, A. R. K.** The Shrewsbury and Newport Canals, *Thomas Telford: Engineer,* Penfold, A., Ed. Telford (1980), pp. 28–29.
37. **Penfold, A.** Managerial organization on the Caledonian Canal. *Thomas Telford: Engineer.* Penfold, A., Ed. Telford (1980), p. 147.
38. **Sandström,** op. cit., pp. 206–207.

10 THE ADVENT OF STEAM AND MECHANICAL ENGINEERING

EARLY HISTORY

The ancient philosophers and men of science in Persia and Greece understood the importance of heat in the scheme of things as shown by their designation of it as one of four basic elements of the universe, i.e., earth, air, fire, and water. The ancient builders and engineers utilized air and water for prime movers only rarely, but it was with the Arabic and Byzantine periods we see their first serious utilization as power sources. With the exception of Hero's *Pneumatics*, the ancient world is silent on heat as a source of work. The true utilization of heat as an energy source came with the development of the steam pump (Savery) and the engine (Newcomen, Watt, Trevithick). The development of this new source of power preceded the theoretical description and analysis of it.

Britain, in the 17th and 18th centuries, laid the groundwork for the Industrial Revolution of the 19th century by maximizing the potential of natural prime movers, i.e., animal, wind, and water power. Animal power was utilized in "horse mills" and "horse-ground gears," with ratios of 100:1 for rotary mill power. Wind power came in the form of post mills, smock and tower mills ("cap mills"), whereby Britons pumped and ground. Post mills were the earliest design, going back to medieval times (Figure 10-1). Tower or "cap" mills were later and only the upper part or cap rotated to meet the wind. A fantail invented in 1750 by Andrew Meikle (1719–1811) maintained the mill's proper heading into the wind and prevented blow-over.

John Smeaton and others (Jean V. Poncelet, 1778–1867) designed water wheels that more efficiently employed water power. The use of "buckets" instead of paddles utilized the force of gravity rather than incident current to make more efficient water-powered mills and blast furnaces. In the U.S. the "pitchback" wheel used the force of the current on the nearside and rotated the lower part of the wheel with the current. It was still inefficient compared to using all of a head of falling water. Poncelet's wheel used a narrow penstock with the current delivered to the offset, curved buckets of the undershot wheel (Figure 10-2, upper). One in 1849 (Spain), 5 meters in diameter, 9 meters wide with only a 2 meter head of falling water, generated 180 hp. The ultimate expression of the water wheel was (and is) the reaction turbine.

Claude Burdin (1790–1873) carried the term "turbine" from the Latin *turbo* ("to spin"). Burdin's student Mercel Crozet Benoit Fourneyman (1802–1867) succeeded in designing the first efficient turbine (Figure 10-2, lower). He used a 106 meter pressure head through a 5 meter penstock or pipe at 160 psi onto a bronze turbine wheel 30 cm diameter (16 kg), spinning at 2300 rpm, yielding 60 hp. Later Fourneyman turbines used only 5 meters of head and developed 220 hp. Compared with the water wheel's efficiency (22%) the Fourneyman turbines (80%) show the tremendous improvement these devices made.

ADVENT OF STEAM AND THE INDUSTRIAL REVOLUTION

We recall Hero of Alexandria's steam turbine (the aeophile) as the first recorded consideration of the use of steam as a prime mover. It was not until the 18th century that ideas concerning the effective use of steam were reified into fact by Thomas Savery (1650–1715). The development of steam as a prime mover was linked to a fundamental problem encountered from classical Antiquity to the Industrial Revolution — dewatering of mines.

As was touched on in the preceding chapter's discussion of canals, the exploitation of wood for fuel and building materials had created an "energy crisis" in the 16th century.[1]

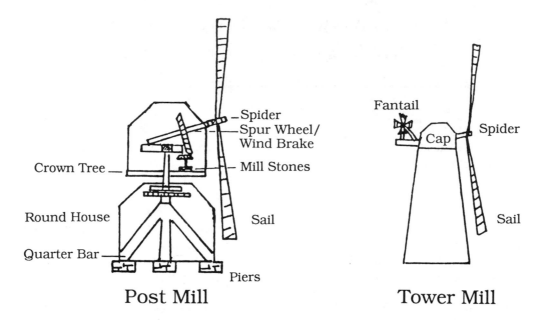

FIGURE 10-1.   Tower and Post Mills.

Deforestation led to the use of coal in Britain, which changed the economic history of that nation and later the rest of the world. As Agricola[2] and other writers[3] have told us, Europe had major mining centers exploiting iron and copper, as well as coal. Britain, including both England and Scotland, sat atop rich coal deposits.[4] Over 3 million tons were mined annually by the 17th century. As the mines went deeper the damage of water increased and dewatering became critical. Horses and overshot water wheels to run pumps were expensive. Horses ate expensive fodder and water wheels required feedstock, which in turn necessitated expenditures on dams for reservoirs. Building these impoundment structures created a whole generation of hydraulic engineers in Britain. More importantly, it led to the development of steam power. By the 19th century the disparity in cost between coal and wood was obvious.

Savery, in 1698, patented a mine drainage pump. This device produced a steam-generated vacuum by spraying cool water on a steam-filled vessel. The resultant partial vacuum raised water in a suction pipe through a non-return valve. Steam pressure was then used to eject the water up a delivery pipe. Savery's device was adequate for moderate lifts and consumed huge amounts of coal.

Savery compared the work with that performed by a horse and invented the term "horsepower". Denis Papin (1647–1712) first proposed that a vacuum could be produced by condensing steam.[5] In 1710 an English blacksmith, Thomas Newcomen, saw one of Savery's engineers and developed an engine with a piston within a cylinder. Completed in 1712, it was employed in mine pumping.

A Newcomen engine is shown in Figure 10-3. The piston rod is connected to the pumps. In operation, the Newcomen engine was little more than a cylinder on top of a boiler. Two jets—one steam, the other water—alternately bathed the cylinder, creating a vacuum allowing atmospheric pressure to force the piston down. Circulation arcs at the beam end let the rods rise and fall in vertical lines. Newcomen engines could lift more water than the Savery pump. This engine practically changed mining overnight. Automatic valves increased the frequency of power cycles to 15/min, e.g., 1739. A Newcomen engine with a 76 cm cylinder and 23 cm stroke (15 strokes/min) lifted water 27 meters. Newcomen engines were very inefficient in terms of the Rankine cycle. Brass cylinders were expensive and created frictional losses. They were not rotary, but rather reciprocating.

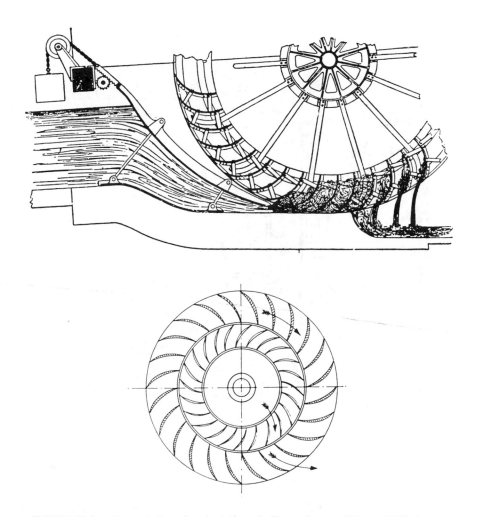

FIGURE 10-2. *Upper:* A Poncelet wheel (from J. Glynn, *Power of Water*, 1852). *Lower:*
Forneyman's turbine.

James Watt (1736–1819), some say, reinvented the steam engine after Savery and
Newcomen. Watt was an engineer, whereas Savery and Newcomen were not. Watt invented
(1) the steam condenser, (2) parallel motion gearing, (3) double-action engines, and (4) the
governor (centrifugal). Matthew Boulton and Watt produced the first rotative design and
standardized 1 hp as equal to 33,000 ft-lbs/min. Oliver Evans (1755–1819) and Richard
Trevithick (1771–1833) pioneered direct-acting, high-pressure steam engines. Evans' "grass-
hopper" engine with offset fulcrum for beam aligned beam with piston rod did not need
parallel motion gearing. Of 500 Boulton and Watt engines in 1800, 38% pumped; 62% were
rotary. Trevithick laid the direct-acting engine horizontal and created the locomotive (Figure
10-4). J.C. Hornblower (1753–1815) invented the compound engine.

Early boilers were inefficient and low pressure. The first boilers were copper with cast
iron heads — wrought iron with rivets was used by 1725. "Haystack" and "wagon" boilers
were geometrically incorrect for steam; cylindrical cones were more stable (Figure 10-5).
Increasing the heating surface led to use of more flues or fire tubes.

It is important to reflect on the meaning of the word "engineer" in relation to the late
18th and 19th centuries. Originally, it had only a military meaning — the man who erected
and worked the intricate engines of war: the catapult, the siege towers, the fortifications or
camps. In 18th century Europe the meaning was expanded and given a civil application —

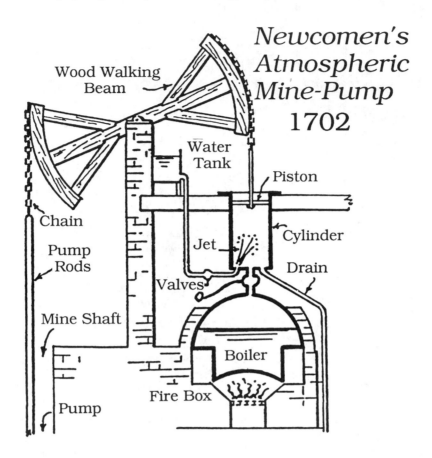

Newcomen's Atmospheric Mine-Pump 1702

Wood Walking Beam

Water Tank

Piston

Chain

Cylinder

Jet

Pump Rods

Drain

Valves

Mine Shaft

Boiler

Fire Box

Pump

FIGURE 10-3. Newcomen's Atmosphere Mine Pump, 1712. (Redrawn from Finch, 1960.)

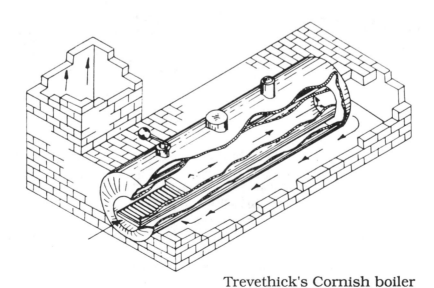

Trevethick's Cornish boiler

FIGURE 10-4. Trevethick's Cornish boiler.

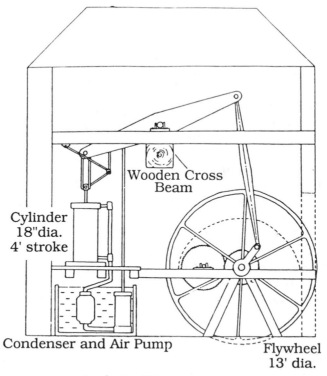

FIGURE 10-5.   Early boiler, fire tube design, ca. 19th century.

builders of canals, bridges and wharves. It was now considered a highly skilled craft rather than a learned profession, men like Smeaton aside. It is now with the advent of steam that we see the beginnings of mechanical engineering, just as chemical engineering can trace its roots to Agricola. The mechanical engineers of the late 18th century — Evans, Savery, Trevithick, Newcomen — "invented" themselves.

CARNOT CYCLE — FRENCH THEORY

The first person to analytically investigate the properties of heat energy was the brilliant French military engineer, Nicolas Leonard Sadi Carnot (1796–1832). Carnot has been justly called the "father of thermodynamics". At age 28 he published the seminal work on the study of heat entitled *Reflexions sur la Puissance Motrice de Feu (Reflections on the Motive Power of Heat)*.

The central thesis of *Reflections* was exposition of the ideal thermodynamic cycle named for him, the Carnot cycle. Carnot demonstrated that the most efficient thermodynamic cycle attainable is that in which all the heat is supplied at a fixed isothermal higher temperature, and all the heat exhausted to the surroundings at a fixed lower temperature. It has been shown that the thermal efficiency of an engine operating upon a Carnot cycle is the ratio of the difference of the higher and lower temperature, to the higher temperature or

$$\text{Thermal efficiency} = T_1 - T_2/T_1$$

All practical engine cycles, Rankine (steam), Otto (gas), Diesel and Stirling (constant pressure), are compared to the Carnot cycle as a basis of comparison where the efficiency of a Carnot engine is

$$\text{Efficiency (Carnot)} = 1 - (T_c/T_H)$$

where T_H is the intake temperature and T_C is the exhaust temperature.

Research has shown that: *No engine operating between two given temperatures can be more efficient than a Carnot engine operating between the same two temperatures.* The Carnot-derived theorem named for him further states: *all Carnot engines operating between the same two temperatures have the same efficiency, irrespective of the nature of the working substance.* It follows that the efficiency of a Carnot engine is independent of the nature of the working substance and is a function only of the temperatures.

Sadi Carnot was born June 1, 1796 in the palace of Luxembourg in Paris. His father, Lazare Nicholas Marguerite Carnot, lived there as a member of the Directory, the post-Revolution government. Lazare Carnot had already established his niche in the history of science as a young officer before the Revolution. He had published a number of important works in mathematics and engineering. During the Revolution his participation led to the top administrative echelons of the new regimes. He ultimately served briefly as Napoleon's minister of war.

Sadi entered the École Polytechnique in 1812. In 1814 he was commissioned in the engineer corps and in 1819 was appointed to the general staff of the corps in Paris. He resigned from the army in 1828 and devoted himself completely to his intellectual and technical pursuits until his death from cholera on August 24, 1832.

His great work, *Reflections,* was an analysis of the factors that determine the production of mechanical work from heat in the steam engine and heat engines in general. Carnot pointed out that ''already the steam engine works our mines, impels our ships, excavates our ports and our rivers, forges iron, fashions wood, grinds grains, spins and weaves our clothes, transports the heaviest burdens, etc. It appears that it must some day serve as a universal motor, and be substituted for animal power, waterfalls and air currents...yet, its theory is very little understood and the attempts to improve it are still directed almost by chance.''[6] Questions raised by Carnot included: how were work and heat related? Was there an upper limit to this work? Was water the best medium to use, or might air, as a gas, or alcohol, as a liquid, provide a more effective medium.

In a heat engine work is obtained by passing heat from a body at a high temperature — a boiler — to another body in which the temperature is lower — the condenser — as in Newcomen's or Watt's earlier designs.

The condition for the production of maximum work was that there should not occur any change of temperature in the bodies employed which was not due to a change of volume, the main offender being direct conduction of heat through the working of the engine. Loss of heat to friction or convection was equivalent to wasting a difference of temperature which could have been utilized to produce useful power. Carnot defined this condition of maximum efficiency as reversibility, i.e., if the reverse of such a process were carried through, all of the effects would be of the same magnitude but reversed in direction.

Carnot was able to demonstrate that the maximum effect was obtained when an engine operated with reversibility and that all reversible engineers operating between the same two temperatures must have the same efficiency, regardless of the medium utilized as the carrier of the heat.[7] The reversible operation of the engine alone determined its efficiency in producing mechanical work. Thus, the greater the relative difference of temperature between boiler and condenser the greater the work for the same difference of temperature; the engine would operate more efficiently, the lower these temperatures. Kerker summarized the principles for heat engines as follows:[8]

1.   Maintain the greatest possible temperature difference between boiler and condenser.
2.   If faced with a choice between equal temperature intervals, select the one lower on the temperature cycle.

3. The actual working medium is unimportant from the point of view of thermodynamic efficiency, except as its properties affect the working temperatures.
4. Seek to operate as closely to the reversible conditions as is practical.

Tragically, Carnot died from cholera at the age of 36 in 1832. Carnot's work is made even more impressive as it was carried out before formulation of the First and Second Laws of Thermodynamics. Carnot's great treatise was little appreciated until work by Clapeyron and Clausius explained the Carnot principle by their famous equation

$$dP/p = (h_{fg}/R)(dT/T^2)$$

where h_{fg} is the heat of evaporation, R is the specific gas constant, P and T, pressure and temperature.

LAWS OF THERMODYNAMICS

The Clausius-Clapeyron equation is the joint result of theoretical work done by a German theoretical physicist, Rudolf Julius Emmanuel Clausius (1822–1888) and the French civil engineer, Benoit Pierre Emile Clapeyron (1799–1864). It is contemporaneous with the work of James Prescott Joule (1818–1889) and the work of all three men resulted in the formulation of the First (Clausius and Clapeyron) and Second Laws of Thermodynamics. Both laws are foreshadowed in Carnot's *Reflections*.

Simply stated, the first two laws are: *when a system undergoes a cyclic change, the net heat to (or from) the system is equal to the net work from (or to) the system* (First Law); or, simply: one form of energy may be converted into another. *No process is possible whose sole result is the absorption of heat from a single reservoir at a single temperature and the conversion of this heat completely into mechanical work* (Second Law).

The second law of thermodynamics tells us that it is impossible to construct a power cycle that operates with 100% efficiency. It also tells us what maximum theoretical efficiency we can expect. The maximum thermal efficiency of the ultimate perfect power cycle is given by

$$n_{max} = 1 \frac{T_L}{T_H}$$

where T_H = maximum absolute temperature of the energy source (where Q_H comes from), and T_L = maximum absolute temperature of the energy sink (where Q_L goes).

Thus, we observe, as Carnot tells us, that the thermal efficiency of power cycles might be improved by increasing T_H, the temperature of the heat source, or decreasing T_L, the temperature of the heat sink.

The statement of the Second Law derives principally from the work of Clausius. Clearly, it derives from Carnot's principles.[9] The essence of the Second Law has been rightly called the "Law of Entropy", or "Law of Organized Energy". The second law of thermodynamics verifies that energy in a highly organized form, available for doing work, is generally more useful and valuable than an equal quantity of disorganized energy.

Clausius, in his paper "Uber die Bewegende Kraft der Warme" (On the Motive Power of Heat"),[10] resolved the problem of expressing calorifically derived Carnot Theory in terms of the mechanical theory of heat[11] by assuming that heat could not by itself pass from a colder to a hotter body. With this resolved, Lord Kelvin formulated the Second Law in the following year, 1851. Clausius expressed entropy as the function s defined as:

$$s = \int \frac{dQ}{dt}$$

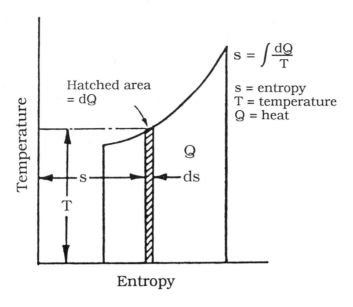

FIGURE 10-6.   Definition of entropy.

Carnot cycle

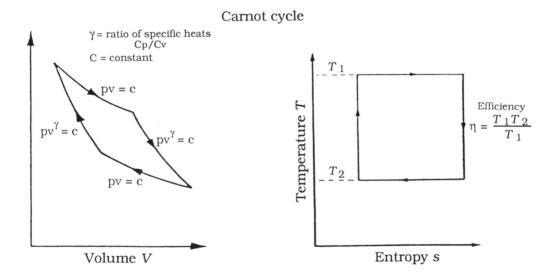

FIGURE 10-7.   *Left:* PV diagram, Carnot cycle. *Right:* Entropy diagrm, Carnot cycle.

where s = entropy, t = temperature and Q = heat.

The practical application of the concept came with the introduction of the Temperature-Entropy or T-S diagram in 1873 (Figure 10-6).

When engine cycles were represented on these diagrams, a more complete picture of the cycle of operation could be obtained by the engineer than was possible with the pressure-volume or pv diagram (see example in Figure 10-7).

As we have seen, the early mechanical engineers designed and built the first steam cycle engines without reference to any elaborate theory on heat engines. As a result, their devices were not very efficient. Newcomen's engines were very costly to operate and were it not for the low cost of coal, the use of these engines may never have been attempted. Watt's engines improved on the efficiency and economy of steam engines but further refinements had to await thinkers like Carnot. Other engineers took different courses to higher efficiency and/or high economy. Two of these men are remembered in engine cycles that bear their names: Otto and Diesel.

Niklaus August Otto (1832–1891) pursued his interests in engineering and science from childhood, even though his parents placed him in commerce at age 16. He continued his studies in his spare time and by 1861 had produced a small four-stroke internal combustion engine. Otto was not the first to think of such an engine, as Alphonse Beau de Rochas took out a patent in 1862 describing the main details of a four-stroke cycle. Beau de Rochas did not include any suggestions as to how it might be used so today it is believed that Otto independently invented the same cycle.

Another claimant to the title of being the inventor of the first internal combustion engine was Jean Joseph Etienne Lenoir (1822–1900). He took out a patent 2 years before Beau de Rochas in 1860.[12] He used illumination gas for his engine and by 1865 had produced over 300 ranging from 1/3 to 3 horsepower. These engines were four-cycle, low in compression, burned fuel poorly and were generally inefficient. Still, Lenoir put one in a carriage and created the first motor car. Later he installed one in a boat. Otto read an article on Lenoir's design and by 1964 had significantly improved on Lenoir's engine. Otto succeeded where Lenoir failed but he must be remembered more for his creativity than his business acumen. By analogy, Lenoir was to Otto as Newcomen was to Watt. Clearly, Otto developed the design into the prototype for all modern gasoline engines.

Otto also described a working two-cycle engine, but this had been invented by Sir Dugald Clerk, born in Glasgow, Scotland in 1854. He first exhibited the two-cycle engine in 1881. Both Otto's and Clerk's engines operated at rotative speeds on the order of 200 rpm. It was for Gottlieb Daimler (1834–1899) to develop the small, relatively high-speed engines for greater power. His engines increased rotative speed to 1000 rpm (compared to 4000–6000 rpm for today's automotive engines). His development of this idea made the automobile a practical reality.

Where Otto's engine took an air and fuel mixture at the suction stroke, Rudolf Diesel (1858–1913) developed one that took in only air, with fuel being injected later at the end of the compression stroke. The burning, theoretically, then proceeds at constant pressure. The remainder of the cycle is the same as in Otto's. The object of this device was to raise the temperature by compression so that the fuel is self-ignited. This allows the Diesel engine to be somewhat more thermally efficient than the internal combustion engine. Diesel engines tend to be slightly more economical in fuel costs as well. Diesel engines are typically used in situations where size and power are not limiting factors. Many diesels today operate on both two-stroke and four-stroke cycles. The two-stroke cycle is popular in motorcycles and small industrial designs as it develops more power than a four-stroke design of the same dimensions.

THERMODYNAMICS — MODERN FORM

Sadi Carnot in 1824 foreshadowed modern thermodynamics with his discussion of the Carnot engine or cycles. He realized the First Law without so stating it — "energy cannot be created or destroyed." The Second Law states that no process is possible whose sole result is the absorption of heat from a reservoir at a single temperature and the conversion of this heat completely into mechanical work. It is the "one-sided law", i.e., "work to heat, but not all heat converted to work." The Third Law of Thermodynamics can be stated as "absolute zero not attainable in thermodynamic processes". This last element of the fundamental laws was the work of William Thompson, Lord Kelvin (1824–1907) who formulated the theory of the conservation of energy and the mechanical equivalent of heat. Carnot had analogized heat transfer from higher to lower temperature to the fall of water over a wheel. This was in line with the prevailing caloric theory of heat. There was no conversion of heat, only a conservation of it.

James Clerk Maxwell (1831–1879), born near Edinburgh, Scotland, wrote on electro magnetic theory and thermodynamics. In the latter he derived the *Maxwell Relations,* which are mathematical relationships essential to the advanced study of properties (of heat engines). These relations involve partial derivatives which consider a variable z which is continuous in x and y or

$$z = f(x,y)$$

thus

$$dz = \left(\frac{\partial z}{\partial x}\right)_y dx + \left(\frac{\partial z}{\partial t}\right) dy$$

It is convenient to express this as

$$dz = Mdx + Ndy$$

where

$$M = \left(\frac{\partial z}{dx}\right)_y$$

or the partial derivative of z with respect to x, with y held constant.

The physical significance of partial derivatives is they relate the properties of pure substances at constant temperature, constant pressure and constant volume. Diagrams of these properties for surfaces whose partial derivatives

$$\left(\frac{\partial z}{\partial V}\right)_T$$

are the slope of the curve, pressure, changing with temperature held constant (constant T process) (Figure 10-8). These surfaces and curves describe the properties of thermodynamic processes and allow prediction of the behavior of thermal devices.

THE SECOND COMING OF IRON AND STEEL

Concomitant with the development of steam, the development of new processes for manufacturing iron and steel were developed in the 19th century. The interaction of an improved iron technology with the developing steam-powered ships and railroads created most of the modern transportation system we use today. Further, the need for coke from coal in iron furnaces after 1856 required steel for the steam engines to pump water out of the coal mines. Before 1856 all iron was produced as wrought iron and cast iron in either small "bloomery" furnaces or charcoal-fueled, masonry blast furnaces. The blast furnace was the indirect-reduction method thought to have been originated by the Romans, and in China.[13] It was used extensively after 1400 in Europe. The end result of the furnace reduction process was wrought iron but it was only possible to produce cast iron. The cast iron, of 2–4% carbon, had to be decarburized in a charcoal-fired hearth called a finery forge. The bloom this produced could be formed, by hammering, in a chafery.

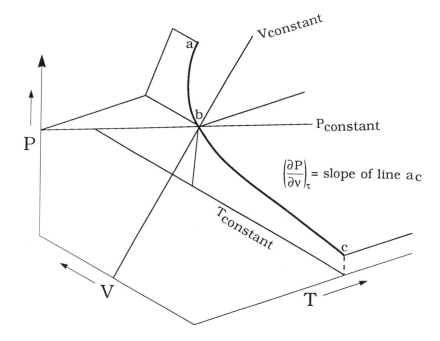

FIGURE 10-8. Pressure-volume surface diagram of partial derivatives.

Iron production on a large scale, utilizing charcoal, required tremendous amounts of wood. In Britain this was a problem due to the competition of shipbuilding needs. In the U.S., less so due to the lower population density and smaller industrial base. Iron production required oil, timber and water to power the blast furnaces. Water was generally available but timber was a critical problem, finally solved in 1709 by Abraham Darby and the use of coke in blast furnaces as fuel.[14] Coke, as a fuel, allowed the iron industry to expand to meet the growing needs of civil, mechanical, and mining engineers. Charcoal was diverted to use in finery forges.

The last vestiges of the iron industry's reliance on charcoal was removed by the invention of the Puddling ("Dry") furnace by Henry Cort (1794).[15] In his manually stirred furnace pig iron was converted to wrought iron. In 1828, J. B. Nelson, a Scot, introduced the hot blast for furnaces, replacing ambient temperatures used previously (Figure 10-9, upper). In 1856 Henry Bessemer decarburized molten iron by blowing cold air through it. As air was bubbled through the white-hot liquid (Figure 10-9, lower), the oxygen of the air and the dissolved carbon combined to raise the furnace temperature to a point where the iron is in a blast of CO_2. Only a fraction of the carbon is left in the iron, which upon cooling becomes a high grade steel (i.e., "mild steel" is a formidable competitor with wrought iron). By 1868 R. R. Musket had introduced tungsten into steel, making a self-hardening metal ideal for engineer's cutting tools, and in 1887 R. A. Hadfield developed manganese steel. In 1913 Harry Brearley developed stainless steel, a direct ancestor of the tremendous range of alloy and special steels used today in aerospace and nuclear applications.

IRON, STEEL AND RAILROADS

Bridges of iron began in Britain in 1779 with the Coalbrookdale Bridge (Figure 10-10). In general, subsequent bridges were linked to the railroads — the bridges spanned streams and bays that blocked the most direct routes for the expanding rail lines.

The scale of 19th century rail construction was impressive. The River Weaver viaduct in 1836 involved the construction of rail line; directed by John Locke, it involved 100 excavations and embankments, whereby 5.5 million cubic meters of earth and stone was

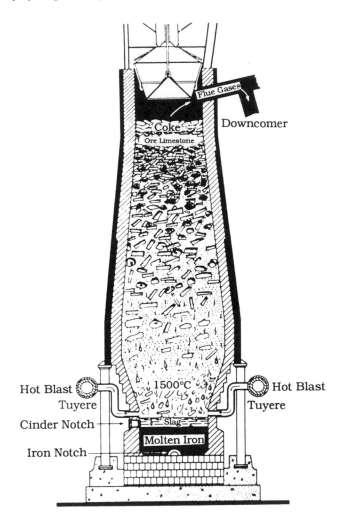

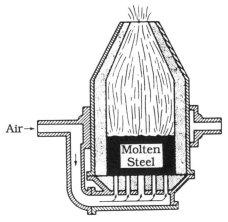

FIGURE 10-9.   *Upper:* A schematic cross-sectional view of an iron blast furnace. *Lower:* A Bessemer converter.

FIGURE 10-10. Coalbrookdale Bridge, 1779 (Library of Congress).

cut and removed. Three million cubic meters was used in the embankments. The railway passed under 100 bridges, two aqueducts, and two tunnels. It passed over 50 bridges and five viaducts between Birmingham and Liverpool.

Richard Trevithick (1771–1833), the originator of high pressure steam engines along with Evans, put these lighter sources of steam power to use in propulsion of vehicles on "rail roads". The development of steam railroads awaited the parallel improvements in producing adequate iron rails. This was possible by 1820.

Early railways had to be nearly level, thus necessitating extensive grade work in the construction of these lines. This is illustrated by the example of the Liverpool line. Robert Stephenson (1803–1859), a colleague of Trevithick's, utilized the combination of smooth steel rails and smooth wheels. Stephenson used Trevithick's device of carrying the boiler flue up a "smoke stack" to develop greater draft. His father, George Stephenson, in 1818 decided that a locomotive must overcome three resistances:

1.   Friction of the axles
2.   Friction between wheel and rail, "rolling resistance"
3.   Gravity on grades

He determined that "friction was constant at all speeds." The effect was thus — with 1:100 an additional pull of 9 kilograms was necessary to overcome gravity or a "1% grade doubled resistance, necessitating three times the tractive effort." The Liverpool-Manchester line was the first decisive success for railroads.

Stephenson's ROCKET (1829) carried a carriage with 30 passengers at 40–50 kph. He used a multitubular boiler (7.5 cm diameter copper tubes) through a stack. The pistons connected directly to the drivers. After the ROCKET the cylinders were moved to the horizontal. Isambard Kingdom Brunel built the Great Western Railway, where he tried a wide gauge of 7 ft (2.1 meters). Stephenson's used a 4 ft. 8½ in. (1.3 meter) standard gauge.[16]

The first rails to be used were cast iron. P. Stevens, son of John Stevens (1749–1838), used wrought iron "T-rails". He understood that stresses of compression and tension greatest at top and bottom, e.g., neutral axis, could be trimmed of excess metal. Other improvements to railroads included:

● wooden cross ties ("sleepers")
● changing radius curves
● swivel trucks (flexed wheelbase)

The Baltimore-Ohio Railroad was America's Liverpool Line, an experimentation that proved successful. However, it was only by the U.S. Civil War that railroads finally overcame steamboats. Railroads, while built across Britain and later the U.S., were not universally loved. One hater of railways said he would rather meet a highwayman, or see a burglar in his own house, than have a surveyor with a theodolite sneaking about his land.

BRIDGES

New bridges of the late 18th and 19th centuries were first built of iron beginning with Abraham Darby's (1750–1791) Iron Bridge at Coalbrookdale. The first modern suspension bridge was erected by J. Finley (1762–1828) in the U.S. in 1801. These were built on the catenary principle. These chain suspension bridges included the Menai Strait Suspension Bridge (579 ft span) by Thomas Telford and the Conway Castle Bridge (416 ft span) built between 1819 and 1829 by Telford as well.

Finley worked empirically to arrive at his bridge by creating a model of a catenary curve with a strong "thread" from which he suspended weights that emulated a loaded cable and determined the position of his bridge's deck.[17] Finley's bridge was built of wrought iron chains with the floor or deck stiffened by wooden trusses. It was 21 meters long, hardly a Menai Strait or Golden Gate edifice, but the genesis was there if not the scale. Because he patented his work in 1808 and published in 1810 he attracted Thomas Pope's critical attention in his *Treatise on Bridge Architecture*.[18] Thomas Telford knew of Finley through Pope by 1818 when he began Menai Strait.

The suspension bridge's cables act like arches in reverse in that they pull on their supports rather than push.[19] The first cable suspension bridge was built in the U.S. in 1816. This suspension design was built as a foot bridge (less than a meter wide) over the Schuylkill River. What is important is the span, which was 124 meters. The bridge fell within a year of its completion, but, as with Finley's bridge, the idea was demonstrated. Better design and materials could succeed in longer-lasting bridges capable of spanning the large distances compared to arch designs.[20]

The early suspension bridge designers knew intuitively that the supporting members of a suspension bridge must be flexible and strong in tension. Telford, at Menai Strait, chose wrought iron eyebars. Wrought iron tensile strength and ductility, unlike cast iron or steel, are greater in the longitudinal direction.[21] Telford assumes a breaking stress of 5625 kilograms per square centimeter (80,000 psi) and multiplied this by the bar's cross-sectional area to arrive at a value for "suspending power" of 12,656 kilograms per square centimeter (180,000 psi).[22] While Telford was heir to the period's general inadequacy in materials science, his approach was good engineering in its imagination and care.

On this rational and careful approach to design, Telford erected one of the most daring suspension bridges over the dramatic setting of the Menai Strait (Figure 10-11). The long central span (176 meters) combined with its height above the waters of the strait made a visual impression on observers of that day and modern day viewers of great suspension designs that is unique to them. The bridge launches its deck on suspension cables that never appear to be equal to the task.

It is with the bold use of suspension design that the 19th century engineer reaches a level that Billington has termed "structural art".[23] After Telford, the French and Swiss engineers applied first principles and new materials to create new designs that made even longer spans. The leaders of this group were Henri Dufour (1787–1875) and five brothers of the Seguin family, of which the oldest was Marc Seguin (1786–1875), 1 year older than Dufour. Seguin wrote a book in 1824 on wire suspension bridges and built the first such design for vehicular traffic.[24] This bridge, built in Geneva in 1822–23 was the beneficiary of Dufour's training at the École Polytechnique and his rigorous experimentation on the strength of wire used in his suspension designs.[25]

Dufour and Seguin practiced in an era of change in Europe. The French Revolution (1789) and the Napoleonic Period which ended with defeat at Waterloo, Belgium in 1815 had forever upset the old ways of doing business politically or otherwise. In Britain's case her triumph led to over 100 years of empire, while France's and other continental monarchies fell or were altered by subsequent revolutions. Dufour was born to Genevan parents in exile at Constance, then part of Germany. He became the unofficial chief engineer for Geneva, the most important city in 19th century Switzerland and perhaps second only to Paris in the French world.[26,27] He was a participant in the wars of Napoleon only in the sense he was posted to the island of Corfu as a military engineer. Dufour returned to Geneva after the wars to serve as head of the Swiss Army twice, work as an engineer, and help establish the International Red Cross.[28]

The impact of the ferment of the early 19th century led to two results in engineering, particularly that of iron construction. Engineers on the continent in countries like France and Switzerland experienced high costs in raw material and manpower.[29,30] Britain, riding the mounting crest of the Industrial Revolution, had access to both cheap and abundant sources of both men and materials. In the case of suspension bridge construction, Dufour's,

FIGURE 10-11.   Menai Strait suspension bridge, 1818–1826. (Library of Congress)

Seguin's and Louis Vicat's (1786–1861) experiments on wire demonstrated its superiority to wrought iron bars or chain.[31,32]

> We can see from this," [concluded Dufour] "the immense advantage of using iron drawn to wires rather than forged into bars: it is more manageable and the force is double. One can proportion the resistance precisely to accommodate the force to be overcome by providing the necessary number of wires, and one can rest assured as to the danger of internal flaws which nothing can show up beforehand in large bars.[33]

Telford and other chain bridge builders (mostly British) countered Dufour's assertion by testing every eyebar or link. To them the cost of a cable (wire) construction was twice that of chain, which easily offset the 137% advantage in strength of wire over chain. Also, they tested so thoroughly because they understood the consequences of the failure in the suspension chain in designs where strong decks were not evolved to aid the structure's overall stability.

It was the Seguins who were responsible for the important introduction of the stiffening truss in their second bridge in 1824. For the Geneva Saint Antoine Bridge they and Dufour followed Brinley's use of a stiff railing and deck. The final form of the first permanent suspension bridge opened on August 1, 1823 with two spans of 40 meters each.[34]

The success of the Saint Antoine bridge led to the 273 meter span of the Sarine River Bridge at Fribourg, Switzerland, built by Joseph Chaley in 1832–34. The "Gran Pont Suspendu" marked the coming of age of the suspension form. It stood until 1923 when it was demolished. Chaley (ca.1800–1870) was trained by Jules Seguin, who used the parallel cable system on a scale not seen before.[35] This system became the standard form after the Sarine Bridge's completion. Chaley's four cables were 13 centimeters ($5\frac{1}{4}$ inches) in diameter and had a strength of 680,400 kilograms (1,500,000 lbs).

A final note on this bridge which stood for 25 years was Dufour's recognition of the property of resonance in suspension bridges.[36] He incorrectly assumed that the sheer mass of the bridge would resist dynamic loads caused by traffic or wind. He was not alone. Engineers of the statue of Navier and Ammann thought only heavy winds could make a heavy bridge swing.[37] They missed the simple point, known to children who swing, that the effect of small push (force) over a long time can be as important as great force over a short time. Once a motion or swing begins less energy is necessary to maintain it or even amplify it, as dramatic failures of suspension designs have shown up to the mid-20th century. It was a collapse of such a bridge, the Basse-Chaîne Bridge, in 1850, with the loss of 226 lives during a wind storm that ended wire-cable bridge construction in France. It was for engineers of the U.S., such as John Roebling, to continue their development.

John Roebling (1806–1869) and his son Washington rivaled Dufour, Seguin, Chaley and the later Eiffel as masters in iron bridge design. Where the latter framed structures in rigid arches, the Roeblings hung them in flexible suspension. The Roeblings reached a peak as engineers with the designs of (1) Niagara Suspension Bridge (1855), (2) Cincinnati (1866), and (3) Brooklyn Bridge (1883).

The Niagara Bridge was the first suspension bridge to carry rail traffic. Its span was 250 meters (821 feet). Lateral stays resisted oscillation and wind loading. It was removed in 1897 as trains became heavier. The Cincinnati Bridge was the prototype for the Brooklyn Bridge. It had one 322 meter (1057 ft) span. The Brooklyn Bridge utilized caisson construction pioneered by Eads in St. Louis. Tensile strength of cables is 160,000 psi with a span of 486 meters (1595.5 feet).

A little-known Roebling-type design of 1866, the Waco Suspension Bridge, was first opened for traffic on January 7, 1870, at which time it was the only bridge across the Brazos River, Texas' longest waterway.

Chartered by the Texas Legislature in 1866 as a private bridge project, work commenced in 1868. The cables and steel work were provided by John A. Roebling and Son, Trenton, New Jersey, who built the Brooklyn Bridge during that same decade. The cost originally estimated at $50,000.00 finally ran to over $144,000.00. In times of high water it was also used extensively by the cattle herds moving up the trail to Kansas. Designed and built for the horse-drawn traffic of that early time, it continued to serve faithfully until the early 1970s.[38]

An early example of Roebling's work is the still extant Delaware and Hudson Canal Aqueduct crossing the Delaware River in New York. It was built in 1847–48 and is the earliest surviving example of a U.S. suspension bridge. It was converted to a highway bridge in 1898.[39]

Another major development in iron bridges was tubular bridges of wrought iron, the most outstanding of which is the Brittania Bridge (Figure 10-12), again over Menai Strait. The Brittania is constructed of two rectangular hollow tubes of wrought iron placed side by side. Each tube carried one of the two tracks of the Chester and Holyhead Railway. The bridge was designed by Robert Stephenson and was completed around 1850. Due to the uncertainty of the design, Stephenson originally planned to add suspension chains but found them unnecessary when he saw how rigid the tube was. This design, which is still in use today, has stood the test of time, fire and the increase in weight and speed of modern trains.

Truss designs were particularly popular in railroad construction (Figure 10-13). Wooden truss bridges were prevalent in the early 19th century U.S., with iron replacing it as a material in truss designs after 1850. Various designs were typically named after their inventors.

In Switzerland, which is considered the birthplace of the timber bridge, Palladio had designed and built several wooden bridges. However, hardly anyone was interested in these wooden bridges at the time. As we have seen, towards the end of the 18th century, advancements were being made in timber spans. Two brothers, Johannes and Hans Ulrich Grubenmann, built three bridges that attained lasting fame, the Schaffhausen Bridge, the Reichenau Bridge, and the Wettingen Bridge. All three of the bridges had certain characteristics in common. They combined the arch and truss principles in their wooden covered bridges, because little was known about how the load was distributed in a truss. Since, by this time, the principle of the arch was well understood, the arch was included in the design.

In 1840 William Howe (Figure 10-13) patented his truss, which was a lattice with vertical iron tension rods. This design was the transition design from the wooden to the iron bridge, and was to become the most popular truss system in America in the close of the 19th century. Thomas Pratt patented a system which was the exact opposite of Howe's truss. The diagonals were of iron and were in tension instead of the verticals (Figure 10-13). The Pratt system was mainly used for iron trusses. The main reasons for the popularity of the Howe and Pratt trusses was the introduction of the railroad and the need to be able to carry heavier loads and much larger spans. Other truss designs include that of F.A.V. Pauli (1802–1883)—a lens-shaped truss-girder, and top chord semi-elliptical truss with diagonals acting as only tie member, by J. W. Schwelder (1823–1894).

Alloy steel tubes were used in building the arch trusses of the Eads Bridge of the Mississippi River at St. Louis. However, in 1855, with the invention of the Bessemer process, a new and more economical material of modern bridges, the use of steel in bridges marked a new era of design.

Steel thus allowed the production of cables with tensile strengths of over 100,000 psi (7000 kg/sq cm). Later cables, such as at Golden Gate (1937), had a tensile strength of 220,000 psi (15,500 kg/sq cm).

The first use of steel in truss construction was made between 1869–1874 with the St. Louis Bridge, several years before the material became low priced. Eads Bridge was in reality a series of four arches, but in construction the cantilever was used to carry the arches over the river; then the truss girder was added to stabilize the arches.

FIGURE 10-12. Robert Stephenson's Britannia Bridge (Library of Congress).

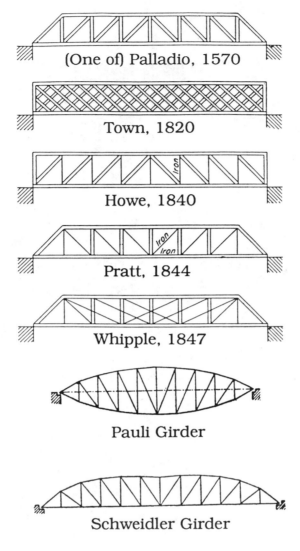

FIGURE 10-13.   Truss designs.

The use of steel was prompted in Britain by one of the worse disasters due to inexperience with cast iron columns — the Tay Bridge disaster in Scotland. After 2 years of apparently satisfactory performance, the bridge suddenly collapsed on December 29, 1879, with a train tumbling into the water causing great loss of life (75 people). That night one of the worst wind storms in history hit and the bridge had not been designed for wind loads. The bridge built to replace the Tay was a cantilever girder design called the Firth of Forth Bridge, built by Benjamin Baker (1840–1907) and John Fowler (1817–1898). Its length is 2530 meters (8300 feet) and center spans (two) are 521 meters (1710 feet) each. Fowler and Baker took care to reduce the wind stresses on the structure. Fowler reduced the vertical cross-section of the 100 meter high towers designing an inward slope from 37 meters at the base to only 10 at the top. The cantilever structures were narrow as well. Baker measured the winds, as well as experimented with models to determine wind loading effects. To his surprise the winds from the southwest were a factor of two greater (20 vs. 40 psi) than those off the North Sea.[40] Baker, like Eads before him, realized that the hollow steel tube was the strongest form by weight. These he fashioned into the great verticals to carry the compressive force

FIGURE 10-14. Eiffel Tower. (Photograph by author.)

of the structure. The Forth Bridge is not a particularly pretty or graceful design bit it is powerful.[41] After the Tay disaster that is what the times called for.

Until the 1850s, Britain, then France, was world leader in steel bridge design. This leadership passed to the U.S. then to Swiss bridge builders by the mid-20th century with their dramatic steel and concrete techniques.

Gustav Eiffel (1832–1923), most famous for the great Paris tower (1899) of his name, was France's leading iron bridge builder. He used the spreading tower design as early as 1869 to resist lateral wind loads and support "trellis" truss girders. He developed the crescent arch form in 1875 with a span of 165 meters between the springings. However great his achievements in iron bridge design, it is by his tower that we remember Eiffel (Figure 10-14).

TABLE 10-1
Eiffel Tower

Number of iron structural components in tower	15,000
Number of rivets	2,5000,000
Weight of foundations	277,602 kg (306 tons)
Weight of iron	7,341,214 kg (8092.2 tons)
Weight of elevator systems	946,000 kg (1042.8 tons)
Total weight	8,564,816 kg (9441 tons)
Pressure on foundations	4.1 to 4.5 kg per sq. cm, depending on pier (58.26 to 64 lbs per sq inch
Height of first platform	57.63 m (189 feet)
Height of second platform	115.73 m (379 ft 8 in)
Height of third platform	276.13 m (905 ft 11 in)
Total height in 1889	300.51 m (985 ft 11 in)
Total height with television antenna	320.755 m (1052 ft 4 in)
Number of steps to the top	1671
Maximum sway at top caused by wind	12 cm (4³/₄ in)
Maximum sway at top caused by metal dilation	18 cm (7 in)
Size of base area	10,281.96 sq m (2.54 acres)
Dates of construction	26 January 1887 to 31 March 1889
Cost of construction	7,799,401.31 francs ($1,505,675.90)

Eiffel built his construction out of iron even though he was aware of the advantages of steel. The tower followed shortly after Eiffel's collaboration with the sculptor Auguste Bartholdi in the design of the Statue of Liberty in New York Harbor.[42] Compared to the Colossus of Rhodes, the design featured Eiffel's iron armature or frame to support the upper metal skin of the statue. The statue thus presaged the Paris tower by only 3 years and followed on a late-19th century fashion of national monuments of colossal dimensions.[43]

In the tower the armature became the sculpture — a carefully designed one. Eiffel took the same care as Baker did with the Forth Bridge as regards wind load. He calculated pressures at all levels of the 300 meter height. The tower is built as a cantilever beam anchored at one end. The legs and tower were built as a tube to even out forces by use of diagonal cross-braces.[44] Table 10-1 lists data on Eiffel's masterpiece.

Nineteenth century Britain produced two unique father-son engineering heritages in the Stephensons and the Brunels. We have already mentioned Robert Stephenson's Brittania Bridge and the ROCKET. Both the Stephensons as well as one of the Brunels, the son, Isambard Kingdom Brunel (1806–1859), were railway engineering pioneers.

The elder Stephenson was self taught, with little formal education. He, like Brindley and others of his generation, solved engineering problems empirically. As often as he was right he invariably was wrong in his engineering judgement. He was a woeful surveyor and intolerant of professionals, which led him to hire assistants as unschooled as himself. The Liverpool and Manchester Railway was badly surveyed and had to be redone. On the Stockton and Darlington Railway, his son Robert, with formal education and apprenticeship under his father to guide him, eliminated the surveying problems. For one thing the line was not built through all obstacles in the way. The Liverpool and Manchester, only 50 kilometers, had a cut, the Olive Mount, that required removal of 800,000 cubic meters of rock, 63 bridges, the Edge Hill Tunnel and crossing the Chat Moss bog.[45] More amazing, perhaps, to us was the construction of such a project without a decision as to what type of tractive power would be used on the line — horses, locomotive, or stationary engine.[46] This led to the famous contest at Rainhill where the youngest Stephenson's ROCKET bested three other designs, one of which was built by John Ericson, by averaging 21 kilometers per hour (13 mph) over a level 2.4 kilometer course. Ericson, Swedish-born, is best remembered as the engineer who designed the USS MONITOR, the revolutionary Civil War ironclad warship.

The Brunel's were heirs to the French engineering tradition. Marc Isambard Brunel

(1769–1849) was a young naval cadet who was tutored by one of the founders of the École Polytechnique, Gaspard Monge (1744–1818).[47] Forced to leave France in 1793, he came to the U.S., becoming Chief Engineer to the city of New York. By 1799 he was in England manufacturing ship's gear for the Royal Navy. He built a sawmill plant at Chatham which still stands today. His son, Isambard Kingdom (the middle name is from his English mother, Sophie Kingdom), was educated at the Lyée Henri-Quatre, graduating in 1822, after being refused entry into the École Polytechnique. This refusal was due in part to his "mixed" parenting, e.g., French-English.

Isambard assisted his father in the latter's greatest challenge, the Thames Tunnel. The 365 meter tunnel was the first subaqueous design. It flooded five times killing seven men, nearly including Isambard, who only barely escaped in one episode in 1828. Marc Brunel drove the tunnel using a square cast iron shield 10.6 by 6.2 meters in size. It was moved forward by hydraulic jacks with workers excavating for a week. It was finally completed in 1843. It was Marc's last project and the apprenticeship of Isambard.[48] In 1833 Isambard was appointed engineer for the Great Western Railway.

Brunel built nine major bridges and viaducts, together with the 2900 meter Box Tunnel, along the 200 kilometer length of the railway. In his career he was to complete almost 2000 kilometer of railroads. He became a master of wood construction, as well as having proposed a whole new gauge for the track (7 feet). This new width he justified as follows: "By widening the rails, the body of the carriage could be kept within the wheels and its center of gravity lowered, producing steadier motion and enabling the wheel diameter to be increased". Brunel found the latter, wheel diameter, to be proportional to frictional resistance.[49] Stephenson's choice of the 4 foot gauge was clearly based on the width of the coal wagonways used at that time.[50] Brunel's "broad gauge" was adopted and larger engines were designed for it by Daniel Gooch (1816–1889). Perhaps even more indicative of the nature of Brunel was his recommendation to "extend" the Great Western all the way to New York, building a regular steamship service.[51] This led him to pioneer in the design of steamships and end the use of sail forever.

Brunel launched GREAT WESTERN, in 1837, as a 2300 ton sidepaddle steamer with 750 hp.[52] In her design he enunciated and applied the broad principles of "economies of scale" in ship size.[53] Up to Brunel's time the common opinion was that given speed increased in proportion to size measured by displacement. This ran counter to Chapman (1775) and Beaufoy (1834) but was generally accepted. Brunel, following Beaufoy, believed that the square-cube law clearly applied to ships such that to increase size meant a more favorable ratio required to fuel capacity. Brunel equated the weight of the ship and its contents to displacement and derived an equation for vessel size still in use today.

Brunel used GREAT WESTERN to make the transition from small to large ship design by recognizing the role that beam-like longitudinal strength was to play in hull design. To Brunel the ship was much like the bridge and that to provide for adequate bending and shear strength meant the design of a hollow box beam much like Stephenson's Brittania Bridge (Figure 10-15).

Brunel went on to design the GREAT BRITAIN (1843) (3443 tons; ave. speed 9.5 knots, 1500 hp, balanced rudder, propeller, bulkheads, clipper bow, with wire rigging, hinged mats). Considered by some to be his finest ship design, she was the first screw-driven, iron ship of any size. Figure 10-16 shows some of her important features.

Brunel then built the GREAT EASTERN, the largest steamship constructed until the 20th century (27,060 tons; 11,000 hp, 692 foot (210 m) length, 83 foot (25 m) beam); it could hold 4000 passengers, 400 crew, 6000–8000 tons of cargo (Figure 10-17). It used longitudinal framing first seen in GREAT BRITAIN and a cellular design in the transverse bulkheads 60 feet (18 m) apart and two longitudinal bulkheads 36 (11 m) feet apart, 350 feet (107 m) long with a double iron hull of 30,000 plates. Using side paddles (56 feet

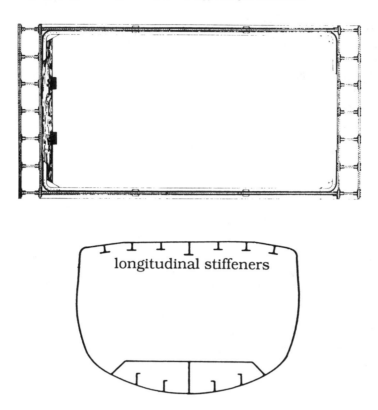

FIGURE 10-15.   *Upper:* A cross-section through the Britannia Tubular Bridge turned through 90 degrees. *Lower:* Simplified structural section through a modern steel ship with longitudinal stiffeners foreshadowed by Brunel in his design. (Redrawn from Beckett, 1980.)

[17 m] diameter) and a screw propeller (24 feet [7.3 m] diameter) to attain 15 knots, it sailed from England to New York in 11 days. GREAT EASTERN brought together two engineers who, along with Brunel, would profoundly shape modern ship design: John Scott Russell and William Froude. Russell's work produced the theory of wave-making and Froude's hydrodynamics.[54] Although GREAT EASTERN was an economic failure, her one success was laying the first Trans-Atlantic cable; she proved that Brunel was correct in believing only iron could create large ships. Further, she was the first large vessel to be designed according to principles of hydrodynamics.

One of Brunel's greatest bridges was the Clifton Suspension Bridge (Figure 10-18), finished after his death (1859) by John Hawkshaw and William H. Barlow in 1864. At Clifton he utilized his French training and his work with his father on two small suspension designs to span 214 meters of the Avon Gorge. Hawkshaw and Barlow replaced Brunel's laminated timber girders with wrought-iron plate girders for stiffening.[55] The bridge opened for traffic in December, 1864 and remains in use today.

Brunel's most unusual design combined in a composite design three classical forms of bridges in one structure — arch, beam, and suspension. This was the Royal Albert Bridge at Saltash (Figure 10-19), a single-track railroad bridge. In its construction he used a caisson to site the central pier. The twin span of the trusses are 139 meters apiece. Each truss weighs 1060 tons. It still carries trains today, as does Stephenson's. Both these great engineers died in 1859 less than one month apart. Stephenson was 56 and Brunel 53. Seldom have any two engineers accomplished so much in a like span.

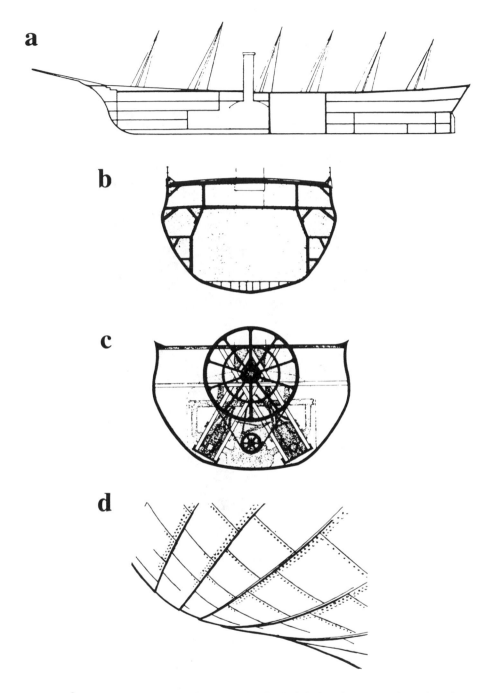

FIGURE 10-16.   S.S. GREAT BRITAIN (a) longitudinal design; (b) cross-section; (c) engine placement; and (d) iron hull plates. (Redrawn from Beckett, 1980.)

NAVAL ENGINEERING

By the mid-18th century, Pierre Bouguer, French mathematician and another founder of modern naval architecture and engineering, developed the concept of metacenter. The distance between M, the metacenter, and G, the center of gravity measure a ship's stability. GM distance is of prime concern in ship design. If G is above M, in narrow hull, high superstructure design, then the ship can capsize. If GM is small, i.e., G is below M but too close, the ship will roll in synchronous motion of wave or swell. If G is too far below

FIGURE 10-17.   GREAT EASTERN (Library of Congress).

FIGURE 10-18. Clifton bridge (Library of Congress).

FIGURE 10-19.   Royal Albert bridge (Library of Congress).

M, the ship will "jerk" upright and be "stiff". A safe GM is approximately 5% of the beam — breadth at widest part of ship. M is defined as the point where B, buoyancy force, intersects the ship's vertical midline.

John W. Griffiths and Donald McKay were designers of "Clippers". Before 1840 ships continued "beamy" designs that hark back to ships of the Middle Ages. Griffith determined shapes of least resistance that follow this law of hydrodynamics — "max. velocity of a solid body moving through a fluid is determined by the radius of the smallest curve along the line of flow." This led to design of the clipper bow, i.e., "turned inside out", on the RAINBOW.

William Froude (1810–1879) worked with Brunel on the design of the GREAT EASTERN using bilge keels to reduce roll. His experience with the GREAT EASTERN convinced him there should be a more rigorous method of evaluating hull performance than with full-sized vessels. Froude practiced as a civil engineer for a number of years and came under the influence of the great bridge and steamship builder I. K. Brunel, who encouraged him to take up naval architecture as an interest. In 1856 Brunel asked Froude to investigate the behavior of the GREAT EASTERN, so Froude carried out a thorough investigation to determine the hull resistance and rolling characteristics on the ship and also on scale models. From this study he produced data which was of great value to naval architects for many years to come.

Froude hoped to derive a formula for hull design based on hull resistance. He determined one source of resistance was the waves made by the ship, e.g., when a boat starts up small waves form at the bow and stern and these get bigger with increasing speed — energy expended in this process is "wave-making resistance".

Froude's towing tank experiments confirmed his theories about waves, i.e., slow-moving models made many small ripples, fast-moving hulls made fewer, larger waves. Froude discovered that equally proportioned, but different size hulls at the same speed, made different wave patterns. But when a large model was speeded up, a velocity was reached at which its wave pattern exactly matched the pattern of a slower-moving small model. Hence, there was a relationship between speed, size and wave pattern for hulls of the same shape.

However, Froude at that point could only measure total resistance. He made an intellectual jump — if another long-recognized source of resistance, friction, were the only other major component of resistance, then measuring it would give wave resistance by simple subtraction, i.e., $R_{Total} = R_{Wave} + R_{Friction}$. Froude abandoned model hulls for a time and used sharp-edged planks, towing them submerged so they made no waves at all. He used various surface coatings and lengths (sand, cotton, paraffin, varnish, etc.). It became apparent that both length and surface quality influenced frictional resistance.

By arduous calculations he derived a method for determining frictional resistance of any plank provided he knew its length, surface area, speed and quality of surface. By subtracting his values for frictional resistance from total resistance, he was able to obtain wave-making resistance. The relationship between models and full-sized hulls was somewhat simple: "resistance (for hulls of the same shape) is directly proportional to displacement at speeds of similar wave pattern."

His final experiment was to make a scale model of a full-sized ship and compare the values obtained with the towed full-size ship. The results were exact. Froude showed that for a true comparison between towing tests on models and tests on full-scale ships the velocity must be proportional to the square root of the length. The dimensionless quantity known as the "Froude Number" is given by $Fr = c/\sqrt{(1\ g)}$, where c is velocity, l is length and g is the acceleration due to gravity. Froude's work is considered one of the greatest engineering advances ever made in ship building and design.

ELECTRICAL ENGINEERING

The marriage of scientific theory and engineering made for rapid evolution of this field during the 19th and 20th centuries. Early experimenters in electricity included, of course,

Benjamin Franklin. Electrostatic attraction of static electricity was discovered using electrostatic "generators" which were simple glass discs turned between felt pads.

Induction of direct current or DC was discovered by Michael Faraday (1791–1867), who deduced the principle of the generator, induction coil, and transformer. Joseph Henry (1797–1878) independently discovered induction (1832). James Maxwell (1831–1879) worked out equations on conductivity, dielectric constants, permeability and force. Andre-Marie Ampere (1775–1836) defined electric potential or pressure as distinguished from current, while George Simon Ohm (1787–1836) defined what we refer to as Ohm's Law, which states one relationship of current to potential and resistance. $E = \rho J$ where J is current density, ρ is potential and E is electric intensity, or Amperes = volts/ohms, or $I = \xi/R$, where ξ equals emf (volts) (difference in electrical potentials analogous to difference in pressures at ends of a pipe). An early use of electricity was Morse's telegraph, which used batteries and relays to increase distance of the code transmission.

Prime movers using steam, water, gas or diesel were necessary for electricity generation. It was known from Faraday's and others work that rotating discs in magnetic field created DC (direct current) and rotating coils created AC (alternating current). Two things were needed for effective use of electricity: (1) cheap power source and (2) need for electrical power. The development of the incandescent bulb (Thomas A. Edison, 1847–1931) and steam-powered generators by Zenobe T. Gramme (1826–1901) allowed for growth of the electrical industry. Edison's bulbs were not the first (earliest 1820) but he was the first to use parallel circuits and 110 volts. Joseph W. Swan (1828–1914) in England developed the first incandescent electrical lamp in 1860. By 1890 AC and DC voltage combined in transmission and 60 cycle frequency were standard in the U.S.[56] By 1900, 110 volts was standard for lighting circuits, this latter done by transformers (capacitors). By 1921, 220,000 volt transmission lines were built. Water and steam turbines generate most electrical power today. Carl Gustaf Patrik de Laval (1845–1913) constructed the first functional steam turbine in 1882, a 75 cm wheel diameter with 40,000 rpm. By 1897 he was using steam and pressure of 3000 psi. Charles Parsons (1854–1931) developed a multistage turbine which drove the generator which produced 7.5 kW at 100 V (7.5 kW = 10 hp). Steam consumption was 130 lbs per kilowatt-hour. Thomas Edison invented the multipole generator. The steam turbines quickly replaced steam-powered generators.

Large-scale supply was derived from the ideas of Sebastion Ziani de Fenati (1864–1930), who argued that electricity generated by large machines was more economical. He also argued for AC because the alternating current transformer provided the most reliable and efficient method of converting the supply to high voltage on leaving the generating station and back to low voltage in sub-stations near to consumers. Fenati designed the Deptford station of 10,000 volts in London in 1887. He used oil-impregnated paper insulation for the first time.

Steam engines, the Williams Engine, coupled to generators operated under the formula, $R = 60f/p$, where R = rpm, f = frequency, and p = the number of poles of the generator. In the use of steam turbines, a single pole generator running at 3000 rpm could replace a 300 rpm, 10 pole generator ala the Williams type. In water-powered turbines, the horsepower available at any time is proportional to 1/19th the product of the head times the flow. Water-power systems, where stream flow is low in summer, became prohibitively expensive where large-storage reservoirs are required. Interconnection of steam plants with water plants allows economical use of both steam power for "base load" (at or near capacity) with water power for "peak loads". By 1921 interconnection was a standard practice. Water power supply peaked at 43% of 84 billion kW hours in 1928. Steam generators have been supplying the bulk of electricity ever since. Water wheels are 90% efficient, where steam turbine plants waste 50–70% of the heat value of the fuel.

In many parts of the world, particularly mid-20th century Texas and the late century Middle East, the slow-speed diesel power station was and is an effective alternative to steam or water turbine. The diesel, in thermal efficiency of over 20% better than gasoline engines, ranks with the best in steam turbine performance. Dr. Rudolph Diesel (1858–1913) invented the diesel, which compresses air sufficiently to ignite fuel injected into the cylinder. Diesel's first engine was built in 1893 with Sulzer of Switzerland acquiring patent rights in 1897. Commercially, Adolphus Busch (St. Louis) manufactured, under American rights a 60 hp, two-cycle Busch-Sulzer Diesel in 1898. Sulzer's first diesel generator plant was built in 1904 at Einsiedeln, Switzerland.

NOTES

1. **Nef, J. U.,** *Civilization,* W. H. Freeman (1979), p. 130.
2. **Agricola, G.,** *De re metallica,* (1556).
3. **Biringuccio, V.,** *De la pirotechnia,* (1540).
4. Nef, op. cit., p. 133.
5. **Kirby et. al.,** *Engineering in History,* McGraw-Hill (1956), p. 156–157.
6. **Carnot, N. L. S.,** *Reflexions sur la Puissance Motrice de Feu* (1824).
7. **Kerker, M.,** "Sadi Carnot and the Steam Engine Engineers." ISIS Vol. 51 (1960), p. 257.
8. Ibid., p. 128.
9. It is interesting to note that Caratheodory in 1909 restated the Second Law without reference to Carnot or any heat engine: "In the vicinity of any particular state 2 of a system, there exist neighboring states 1 that are inaccessible via any adiabatic change from state 2."
10. **Clausius, R. J. E.,** *Abhandlungen über die mechanische Wärmtheorie,* (1864–1867). Trans. W. R. Browne as *Mechanical Theory of Heat,* Macmillian (1879).
11. Joule derived the mechanical equivalent of heat and his name was given to the standard unit which was equal to 778 ft-lbs/BTU.
12. **Carvill, J.,** *Famous Names in Engineering,* Butterworths (1981), p. 44.
13. **Forbes, R. J.,** *Studies in Ancient Technology,* 2nd ed., Vol. 8, Brill (1964) p. 205.
14. **Walker, D.,** The railway engineers, *Great Engineers,* Walker, D., Ed. Academy-St. Martins (1987), p. 111.
15. Ibid.
16. Railroad gauges are reported in their original measurement units.
17. **Finley, J.,** A Description of the Patent Chain Bridge, *The Portfolio,* Vol. 3 (1810), p. 442.
18. **Pope, T.,** *A Treatise on Bridge Architecture,* A. Niven (1811).
19. **Mock, E. B.,** *The Architecture of Bridges,* Museum of Modern Art (1949), p. 11.
20. **Paxton, R. A.,** Menai Bridge, 1818–26, *Thomas Telford: Engineer,* Thomas Telford, Ltd. (1980), p. 84.
21. **Aston, J. and Story, E. B.,** *Wrought Iron,* A. M. Byers (1957), p. 49.
22. **Paxton,** op. cit., p. 85.
23. **Billington, D. P.,** Bridges and the new art of structural engineering, *Am. Sci.,* 72(1): p. 22, (1984).
24. **Seguin, M.,** *Des Ponts en Fil de Fer par Seguin Ainé d'Annonay,* Bachelier (1824).
25. **Peters, T. F.,** *Transitions in Engineering,* Birkhäuser p. 61–93, (1987).
26. Ibid.,
27. **Herold, J. C.,** *The Swiss Without Halos,* Greenwood Press (1948).
28. **Peters,** op. cit., p. 59–62.
29. **Paxton,** op. cit.
30. **Peters,** op. cit., p. 88.
31. Ibid.
32. **Vicat, L. J.,** *Observations Diverses sur la Force et la Durée des Câbles en Fil de Fer.* APC (1836), 1st Sem. p. 207–213.
33. **Peters,** op. cit.
34. **Straub,** op. cit., p. 176.
35. **Peters,** op. cit., p. 94.
36. Ibid., P. 96.
37. Ibid., p. 156.
38. *The Handbook of Waco and McLennan County,* Kelley, D., Ed., p. 286.
39. **Sackheim, D. E.,** *Historic American Engineering Record Catalog,* U.S. Government Printing Office (1976), p. 110.

40. **Kirby** et al., op. cit., p. 313.
41. **Mock,** op. cit., p. 77.
42. **Deswarte, S. and Lamoine, B.,** *L'Architecture et les Ingénieurs,* Moniteur (1979), p. 217.
43. Ibid.
44. **Mark,** op. cit.
45. **Sandström,** op. cit., p. 213.
46. **Kirby et al.,** op. cit., p. 278.
47. **Beckett, D.,** *Brunel's Britain,* David S. Charles (1980), p. 19.
48. **Harding, H.,** Tunnels, *The Works of Isambard Kingdom Brunel,* Pugsley, A., Ed. Cambridge (1976), p. 38.
49. **Beckett,** op. cit., p. 41.
50. Ibid., p. 35.
51. **Kemp, P.,** *The History of Ships,* Longmeadow (1988), p. 150.
52. **Caldwell, J. B.,** The great ships, *The Works of Isambard Kingdom Brunel,* Pugsley, A., Ed. Cambridge (1976), p. 140.
53. Ibid.
54. **Kemp,** op. cit., p. 168.
55. **Beckett,** op. cit., p. 106.
56. **AC** generators were used to raise voltage to a transmission level and then transformed to DC for transmission. After transmission, the voltage was again transformed to AC for use.

11 SANITARY AND HYDRAULIC ENGINEERING

The adequate supply of water for drinking and sewage disposal has been a concern of civilization since its inception. Archaeological evidence from the second millennium B.C. indicate the ancient Indus Valley cities of Harappa and Mohenjo-daro had water supply and sewage disposal systems. The extensive water supplies of Rome and its sister Imperial cities had the objective of providing adequate drinking water and sewage disposal. How successful many of these ancient water systems were can only be guessed. What is sure is that the public authorities of those times were aware of the need for engineered facilities on a large scale to address water supply and sewage demands.

In Asia Minor the evidence of narrow streets and small rooms in houses huddled within defensible walls tells us that crowding in cities like Babylon and Troy was extreme. Garbage accumulated in the houses, where dirt floors were continually being raised by the debris. Human wastes were typically deposited in the nearest street. The water supply, generally open, was from wells, rivers and canals and was prone to pollution. Life expectancy was short, with a high infant mortality.

In Greece, cities such as Samos and Athens built impressive aqueducts (cf. our discussion on Greek engineering). The aqueduct of Pisistratus at Athens runs underground much of its way. Further, water purity was aided by large settling basins where dirt and sediment would sink to the bottom. Water was not supplied to private homes and came principally from public springs and fountains. Sometimes these were in ornate buildings like the Enneakrounos at Athens ("fountain with nine spouts").

The cities supervised water supply and entered into contracts with major users. Pollution of the city water supply could merit the death penalty. Laws in Athens and other cities required that waste matter be carried outside the walls for a certain distance before it was dumped. Drainage sewers were often covered channels and any excess water in the city supply might be used to flush them. The main sewer at Athens drained into a reservoir outside the city gate *(Dipylon)* and the waste water was conducted through a series of canals to enrich the fields near the city. This practice continued throughout European history and up to the 19th century in England.

In Rome the purity of water in aqueducts was maintained by covered channels, reservoirs and settling basins much like those of the Greeks, but on a much greater scale. There were strict laws against pollution and water theft. A water commissioner, the *aqualegus*, oversaw the vast public water supply of ancient Rome at its height. Noted authors like Vitruvius warned against lead pipes but this warning went unheeded. Some aqueducts carried water unfit for drinking and were used for other purposes. Romans recognized the temperature and taste of water from various aqueducts and had definite preferences. Water from the aqueducts were distributed to public fountains, industries (textiles), some private houses and public baths, which used prodigious amounts of water. The reservoir of the Baths of Caracalla alone contained 76,124 cubic meters.

Furnaces, where hot springs were not available, raised water to the desired temperatures and heated the floors and walls of the baths *(hypocausts)*. Excess water from the aqueducts was used to remove sewage via the Cloaca Maxima. This large sewer system also doubled as a storm drain system. The Cloaca Maxima was 4.3 meters high and 3.2 meters wide in places. There was no sewage treatment, with waste discharged into the Tiber River and thence to the Mediterranean Sea. Floods often backed up into these drains. Public latrines were maintained in Rome.

As we have seen, the majority of Romans lived in *insulae*, large apartment houses which were often too tall for their foundations and supports. Augustus finally set a maximum height limit of 20 meters for private buildings. No running water was inside these insulae. Public latrines (most used interior commodes) were available in some insulae.

MIDDLE AGES

With the fall of the Western Roman Empire, a lack of central authority and large public funds led to a decline in public services such as road building and water supply. The infrastructure of the Roman (Catholic) Church was responsible for some water supply in cities of Roman date. New cities in Western Europe depended on wells, springs and rivers. Many municipal laws forbade contamination of rivers, but this did not work. Further, wells were generally too close to cesspools and latrines. Seepage and overflow contaminated water supplies and epidemics killed thousands in the Middle Ages, mostly in towns and cities of Western Europe. As a result of such ills as the Black Death, England passed the first Urban Sanitary Act of 1388 which forbade throwing filth and garbage in ditches, rivers and waters.

"Water carriers" distributed water from public fountains and wells. These men provided water to citizens of Paris for a fee in the 12th century. In 1190 King Philipe Auguste built walls about Paris. The abbey at Saint-Laurent had a reservoir at Pré Saint-Gervais, which was fed by springs at Romainville and this water was piped to Paris in lead pipes. The abbey of Saint-Martin des Champs repaired 1200 meters of the old Bellville aqueduct. Mains were laid for public fountains in Paris. In 1404, Charles III ordered all pollution of the Seine to cease. Further repairs of the Bellville aqueduct were made by Parisian merchants in 1457.

The advent of pumps in the 15th and 16th centuries changed the picture for public water supply. Wooden pipes were preferred in many parts of Western Europe. In the 19th century in Augsburg (Germany) 3300 meters of the water mains were still wooden. These pipes were 7 meters long and 10–14 centimeters in diameter. Pressures of 3.5 atm were common. Pottery and lead pipes became less common. Lead pipes of 6.25 centimeters were soldered with tin solder. Cast iron pipes were tried in Augsburg in 1412, but were changed to wooden pipes 4 years later. It is important to reflect on the hydraulic engineering of Rome; it took later civilizations over 1400 years to match its accomplishments in this area of engineering. This began to occur in the 19th century in Europe and America.

18TH–19TH CENTURIES

In England Edwin Chadwick (1800–1890) was the "prime mover" in insisting proper sanitation was an essential prerequisite to any improvement in living conditions—" . . . applications of the science of engineering, of which the medical men know nothing; and to gain power for the applications, and to deal with local rights which stand in the way of practical improvements, some jurisprudence is necessary of which engineers know nothing". He insisted on constant-supply systems. He demonstrated the relationship between unsanitary conditions, defective drainage, overcrowded housing and disease.

The first clear proof that public water supplies could be a source of infection for humans was based on careful epidemiological studies of cholera in the city of London by Dr. John Snow in 1854. Although Snow's study of the contaminated Broad Street pump is the most famous, his definitive work concerned the spread of cholera through water supplied by the Southward and Vauxhall Company and the Lambeth Company. The former obtained its water from the Thames at Battersea, in the middle of London in an area almost certainly polluted with sewage, whereas the Lambeth Company obtained its water considerably up-stream on the Thames, above the major sources of pollution. In one particular area served by these two companies, containing about 300,000 residents, the pipes of both companies were laid in the streets, and houses were connected to one or the other sources of supply. Snow's examination of the statistics of cholera deaths gave striking results. Those houses served by the Lambeth Company had a low incidence of cholera, lower than the average population of London as a whole, whereas those served by the Southward and Vauxhall Company had a very high incidence. As the socioeconomic conditions, climate, soil and all other factors were identical for the populations served by the two companies, Snow concluded that the water supply was transmitting the cholera agent. Snow's study,[1] a classic in the field of epidemiology, is even more impressive when it is realized that at the time he was working, the germ theory of disease had not yet been established.

During the 17th to the early 19th century, a number of improvements in water supply were made, most of these related to improvements in filtration to remove the turbidity of waters. During this same period, the germ theory of disease became firmly established as a result of research by Louis Pasteur (smallpox), Robert Koch, K. J. Eberth (typhoid) and others, and in 1884 Koch isolated the casual agent of cholera, *Vibrio cholera*.

IMPORTANCE OF WATER FILTRATION

The Chelsea Waterworks (London) dates from 1722. Water was stored in a reservoir with a tide mill used to drive pumps. The tide mill worked on the simple principle of a turbine wheel in a race fed by penstock. In this case the water was stockpiled during high tide and discharged to the mill at the ebb. Cyclical, but with a sure source of power, it could pump 20 meters of hydrostatic head, from Thames to Hyde Park.

In the 1740s atmospheric engines were used, with tide mill as supplements as late as 1775. These were replaced by two Watt engines, the first in 1778 and the second in 1803. The second engine had 43.2 hp and pumped 5 cu m/min. It consumed 3.3 kg (of coal)/hp/h.

James Simpson (1799–1869), who grew up at the Chelsea Plant, designed the ''slow-sand filter'' of one half square hectare. ''Constructed by laying down of earthenware drain pipes pierced with holes; first, a layer of coarse gravel...; secondly, a layer of fine gravel; thirdly, a layer of sea shells...; fourthly, a layer of coarse sand; and, fifthly, a layer of fine sand''. All these layers were undulated.

In 1892 a study of cholera by Koch in the German cities of Hamburg and Altona provided some of the best evidence of the importance of water filtration for protection against this disease. The cities of Hamburg and Altona both received their drinking water from the Elbe River, but Altona used filtration, since its water was taken from the Elbe below the city of Hamburg and hence was more grossly contaminated. Hamburg and Altona are contiguous cities, and in some places the border between the two follows a contorted course. Koch traced the incidence of cholera in the 1892 epidemic through these two cities, with special attention directed to the contiguous areas. In such areas it was assumed that climate, soil and other factors would be identical, with the principal variable being the source of water. The results of this study were clear-cut: Altona, even with an inferior water source, had a markedly lower incidence of cholera than Hamburg. Since by this time it was well established that cholera was caused by intestinal bacteria excreted in large numbers in the feces, it was concluded that the role of filtration was to remove the contaminating bacteria from the water.

In the U.S. cholera was not a problem after the mid-19th century; the waterborne disease of particular concern was typhoid fever. In England William Budd had shown by the mid-19th century that typhoid fever was a contagious disease. The causal agent was isolated and identified by Eberth in 1880 and Gaffky in 1884. Although the causal agent, now called *Salmonella typhi*, is transmitted in a variety of ways, one of the most significant is by drinking water.

The first public supply in the U.S. was in St. Augustine (Florida) in 1680. Early colonial water supply and sanitary problems are typified by those of a city nearby St. Augustine — Savannah, GA. Founded in 1733, Savannah occupied a well-drained site on a high bluff overlooking the Savannah River. On the south, east and west were swamps and tidal streams.

Yellow fever, typhoid and other insect and waterborne diseases devastated Savannah in the 18th and 19th centuries. Savannah drew her water from the Savannah River. All canals built to drain her surrounding swamps were graded for a southerly flow. Privy vaults were utilized (4000) until 1902 when sewer drainage was completed. Wells were drilled for water supply. River outlet for sewage proposed in 1897 was adopted. Flow direction in canals was changed to northerly (1910) and sewage treatment (1930s) was begun on outfall of canals — bricked over in most places, open still along former city boundaries. Savannah is typical of most 19th century American cities.

Experiments on water filtration were carried out in the U.S. during the late 1880s and early 1890s, notably by the Massachusetts State Board of Health experiment station established in 1887 at the city of Lawrence. At this station, the treatment of water as well as sewage was considered by an interdisciplinary group that included engineers, chemists and biologists. A leader in this work was W. T. Sedgwick, a professor at the Massachusetts Institute of Technology (MIT), and MIT's influence on water supply research remained strong throughout the first quarter of the 20th century. One important technological advance which made water filtration adaptable to even turbid sources of water was the use of chemical/coagulation filtration processes, patented about 1884 by the brothers, J. W. and I. S. Hyatt.

While the Lawrence experiments were going on, an epidemic of typhoid swept through the city, hitting especially hard at those parts that were using the Merrimac River as its water supply. As a result, the city of Lawrence built a sand filter, and its use led to a marked reduction in the typhoid fever incidence. As reported by Allen Hazen in 1907, the death rate from typhoid fever in Lawrence dropped 79% when the 5 year periods before and after the introduction of the filter were compared. Of additional interest was a reduction in the general death rate (all causes) of 10% from 22.4 to 19.9 per 1000 living.

Another major series of filtration experiments was made in 1895–1907 at Louisville, KY, where the source of water was the muddy, polluted Ohio River. These experiments were successful, and from an engineering point of view were of importance because they showed that it was possible to treat source waters of a rather poor quality (the Merrimac River at Lawrence may have been polluted, but at least it was a clear water, making filtration rather easier). The success of the Louisville experiments and the other studies led to rapid establishment of filters as a means of water purification; by 1907[2] Hazen could list 33 cities in the U.S., some of comparatively large size, which were using mechanical filters, and 13 cities that were using slow sand filters. Filtration led to an elimination of turbidity and color from the water, and to a removal of about 99% of the bacteria present. At that time these conditions were considered as a standard by which the quality of a treated water should be judged. As Hazen states: "There is no final reason for such standards. They have been adopted by consent because they represent a purification that is reasonably satisfactory and that can be reached at a cost which is not burdensome to those who have to pay for it...There is no evidence that the germs (characteristic of sewage pollution) so left in the water are in any way injurious. Certainly if injurious influence is exercised, it is too small to be determined or measured by any methods now at our disposal." This last statement is of considerable importance when considered in light of the important advance in water purification practice yet to come, chlorination.

An excellent overview of the relationship between water quality and typhoid fever incidence was published at about this time by Fuertes. He gathered typhoid fever statistics for a large number of cities in North America and Europe, and grouped the data by type of source water and water treatment.

CHLORINATION — THE MOST SIGNIFICANT ADVANCE IN WATER TREATMENT

Although a reading of Hazen in 1907 might lead one to conclude that excellent water quality had been well established by filtration, the most important technological advance in water treatment was yet to come. The introduction of chlorination after 1908 provided a cheap, reproducible method of insuring the bacteriological quality of water. Chlorination has come down to us today as one of the major factors insuring safety of our drinking water. Calcium hypochlorite was manufactured industrially for use as a bleaching powder and was used in paper mills and textile industries. It was a cheap chemical and, hence, readily adaptable to use on the large scale necessary for drinking water. The first practical demonstration in the U.S. of its use in water supply was at the filter plant of the Chicago Stockyards, where it was introduced by G. A. Johnson in the fall of 1908.

The use of chlorination in an urban water supply was introduced in Jersey City, NJ, in the latter part of 1908. The circumstances surrounding the Jersey City case are of some interest from a historical point of view and will be reviewed briefly. Jersey City received its water from a private company which used a large reservoir at Boonton, an impoundment of the Rockaway River. The water was supplied to the city unfiltered, although some settling took place in the reservoir. Several years before 1908, the city raised the contention that the water being supplied was not pure and wholesome for drinking at all times, as required by the terms of its contract with the private company. At certain times of the year the water in the reservoir became polluted as a result of sewage influx from communities on the river above the reservoir. Rather than undergo the expense of a filtration plant or attempt to control the sewage influx from the various communities, the private company chose to introduce a chlorination system. The results were dramatic: a marked drop in total bacterial count was obtained, and at a cost far lower than for any other procedure. Further testimony before the court was held, after many months of operation, to determine whether or not the company was meeting the terms of its contract. The court decided that the evidence was favorable to the company. As stated by the court examiner: "I do therefore find and report that this device (chlorination) is capable of rendering the water delivered to Jersey City pure and wholesome for the purposes for which it is intended and is effective in removing from the water those dangerous germs which were deemed by the decree to possibly exist therein at certain times."

The dramatic effect that chlorination had on water supply problems is well illustrated by comparing the first and second editions of Hazen's book (1907 and 1914). In the first edition, barely any mention of disinfection is made (merely a remark about ozone being too expensive), but in the second edition Hazen waxes enthusiastic about the advantages of chlorination. As he says, chlorination could be used "at a cost so low where it was required or advantageous...When the advantages to be obtained by this simple and inexpensive treatment became realized, as a result of the publicity given by the Jersey City experience, the use of the process extended with unprecedented rapidity, until at the present (1914) the greater part of the water supplied in cities in the United States is treated in this way or by some substitute and equivalent method".

Interestingly, the introduction of chlorination also markedly changed the established ideas about water quality standards: "The use of methods of disinfection has changed these standards radically. By their use it has been found possible to remove most of the remaining bacteria so that the water supplied can be as easily and certainly held within one-tenth of one percent of those in the raw water, as it formerly could be held within one percent...Even today the limit has not been reached. It may be admitted that the time will come when a still higher degree of bacterial efficiency will be required. Present conditions do not seem to demand it, but we must expect that in some time in the future conditions will arise which will make it necessary. When additional purification is required it can be furnished".

Allen Hazen's (1869–1930) importance lies in the fact that he was a major consulting engineer for a wide variety of water works, and was very influential in recommending treatment methods. Chlorination was introduced at about the time that adequate methods of bacteriological examination of water had been developed, thereby permitting an objective evaluation of the efficiency of treatment. This evaluation was not based on the incidence of typhoid fever directly, but on an indirect evaluation using bacterial or coliform counts.

Soon after the introduction of chlorination, it became possible to obtain firm epidemi-

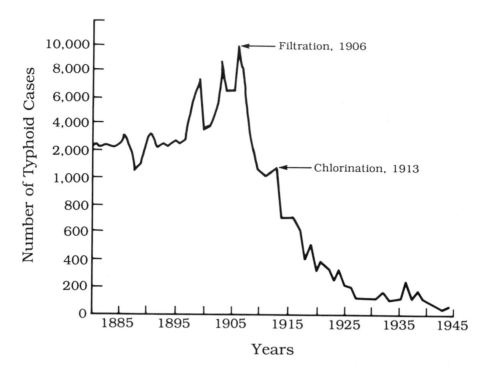

FIGURE 11-1.   Number of typhoid cases.

ological evidence that cities who were chlorinating water had lowered incidences of typhoid fever. The incidence of typhoid fever in Philadelphia during the years 1880 and 1945 is shown in Figure 11-1. Filtration was introduced in 1906 and chlorination in 1913; both led to marked reductions in the incidence of typhoid fever. Another dramatic example is derived from observations at Wheeling, WV, in 1917–1918. The incidence of typhoid fever in Wheeling was 11–200 per 100,000 during these years. Chlorination was introduced in the latter part of 1918, with the result that only seven cases were recorded during the first 3 months of 1919. Chlorination was discontinued for 3 weeks during April 1919, with a resulting increase to 21 cases, or an increase of 300%. Thereafter, chlorination was continued and only 11 cases were recorded for the last 6 months of the year. Other examples of this sort could be cited. As local water supplies (wells, rivers) became inadequate or polluted, other engineers like Hazen were called upon — in the U.S. as in Britain — to impound water for distribution.

DAMS AND WATER SUPPLY

Dams, generally earthen embankments, were built to hold reservoirs throughout England in the mid- to late-19th century. The cream of a generation of English civil engineers worked on major waterworks schemes with reservoirs, aqueducts and embankment dams. While not as well remembered as the Stephensons or Brunels, they produced works, nonetheless, that, in the case of their dams, are still in use after 100 years. Over 90 major[3] dams were built in this period by engineers such as George Leather (1787–1870), his son, John Wignall Leather (1810–1887), James Simpson (1799–1869), James Leslie (1801–1889), Thomas Hawksley (1807–1892) and J. F. LaTrobe Bateman (1810–1889). Bateman and his office built, by far and away, the most with a total of 43 major dams. Only two of the constructions of this era failed catastrophically — George Leather's Bilberry dam (1852) and the Dale Dyke dam (1864) built, ironically, by John Towlerton Leather. The latter was nephew to the former. John Leather's reputation survived the collapse of the Dale Dyke dam, while George Leather's brilliant career came to a tragic end in obscurity after the Bilberry disaster.

In the U.S. water supply projects produced similar successes and failures. The American West provides two examples that illustrate both — the Hoover and St. Francis dams, while the American East provides illustrations of hydraulic engineering for water supply and water transportation — the Croton Aqueduct and the Erie Canal.

By the early 1900s, Los Angeles was a pueblo turned into a boomtown. It had outgrown a water supply composed of groundwater wells and the Los Angeles River. In 1924 Los Angeles began construction of the Los Angeles-Owens Valley Aqueduct, which featured construction of the St. Francis dam. Unlike the embankment dams of the Leathers, Batemans and Hawksleys, the St. Francis dam was a 200-foot high concrete and steel structure that impounded 38,168 acre feet of water. It was designed by William Mulholland, general manager and chief engineer for the Los Angeles Department of Water and Power.

Mulholland (1855–1935) was born in Ireland, coming to the U.S. when he was 17 years old. In 1877 he began to work as a laborer for the Los Angeles Water company. By reading every available book and technical journal and by working with experts, he became a self-taught hydraulic engineer. By 1913 he was lionized for the successful completion of the Owens Valley and Los Angeles Aqueduct, which was hailed as an engineering triumph on the scale of the Panama Canal.

Built as part of later additions to this water supply system, St. Francis dam was completed March 1, 1926. It collapsed on March 12, 1928, leaving in its wake a death toll of over 450 — second only to the San Francisco earthquake of 1906 in terms of property and lives lost in California history.[4] It was never rebuilt and its failure ended the distinguished career of Mulholland just as the Bilberry disaster ended that of George Leather.

Southern California's need for an ever-larger water supply moved beyond the St. Francis disaster to create the 389 kilometers long Colorado River Aqueduct and its centerpeice — Hoover Dam. Authorized as the Boulder Canyon Project in 1928, the $165 million dam and power plant were completed in 1935. The giant power facility was necessary, as 36% of its output pumps water through the Colorado River Aqueduct. At completion, the dam's crest was 220 meters high, forming Lake Mead, the largest man-made lake in America (35 billion cubic meters). The massive construction pioneered mass concrete emplacement and cooling by dividing the dam into 230 giant sections or blocks and embedding cold water flow pipes within them. These 2.5 centimeters pipes allowed the concrete to lose the great heat generated by the curing process in 2 months rather than in half a century or more. Injection of a water and cement mixture between the blocks created the monolithic structure.

In the eastern U.S. the New York City Water Supply is a typical example of the need for extensive hydraulic engineering, matching and surpassing that of antiquity, with 40 kilometers of bored log mains that supplied 2.6 million liters to 2000 homes in the early 1800s (60,000 population).

As needs grew, the Croton Aqueduct (Figure 11-2) was completed in 1842 with 830 liters/capita for a population of 360,000. The slope was 33 centimeters to the mile. A new aqueduct was built in 1885–1893, with a cross-section of 90 meters. The Harlem River Valley was crossed by a pressure tunnel (a la Pergamon) 300 feet deep and 11.2 kilometers long, dug at 7–12 square meters/week and by 1905, the tallest dam. Stone masonry was completed on Croton River with a crest 90 meters high. It stored 380 billion liters of water.

The Catskill water supply system, begun in 1907, was finished in 1937. On it a 76.8 meter high Olive Bridge dam was built of concrete with embedded large boulders, termed *cyclopean* masonry construction. The Catskill Aqueduct increased in cross-section to 23 square meters and featured a 900 meter long pressure tunnel under the Hudson River, 335 meters below the river; 4.3 meter diameter lined with plain, nonreinforced concrete (35 centimeters thick). The water pressure was 42 kg/cm^2 under gravity flow. The Delaware Aqueduct, in places 90–300 meters below ground surface, was dug at 42–82 meters/week and is 170 kilometers in length.

Public water supplies assumed added importance with the progressive increase in ur-

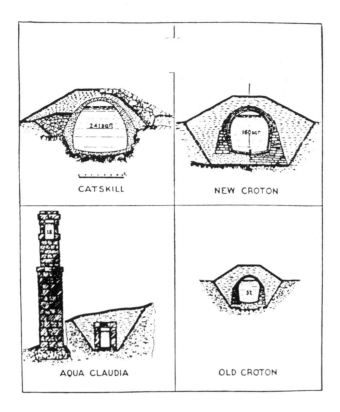

CATSKILL   NEW CROTON

AQUA CLAUDIA   OLD CROTON

FIGURE 11-2.   Croton aqueduct.

banization. Hydraulic engineers were called upon to create water courses for commerce as well. As a water supply, these channels are aqueducts; as routes of transport, they are canals.

SUEZ, ERIE AND PANAMA CANALS

Suez — This construction reduced the travel distance from Liverpool to Bombay from 17,000 to 10,000 kilometers and was built from 1859–1869. It was designed by Ferdinand Marie de Lesseps and called a "ditch-digger's dream". Extending 163 kilometers, it had no locks and averaged 8 meters in depth. Begun in 1859 and finished in 1869, the Suez Canal connects the Mediterranean and Red Seas across 160 kilometers of the Isthmus of Suez.

The canal is a relatively straight cut from Port Said on the Mediterranean to Ismailia. From there it follows a series of shallow lakes to the Red Sea. The canal crosses no greater height than 15 meters at any point so it was just as it was styled — a long, shallow (8 meters) ditch. It has no elaborate locks or control structures. In 1875 it was deepened 3 meters and widened from 70 meters by the British, to become the commercially viable waterway it is today.

Sandström[5] tells the story of its financing, construction and imbroglios with gusto. Its originator, Ferdinand de Lesseps, was no engineer. He was at best an unprincipled entrepreneur who managed the canal project from 1859 to 1863 using corveé labor, including children. After his removal from actual management, contractors used steam dredging equipment to complete the work. Even with 8000 paid workers the project was plagued by debt and outbreaks of cholera which killed hundreds. Still these casualties were trivial compared to the over 40,000 who died from yellow fever in an abortive attempt by de Lesseps to build a canal through the jungles of Panama in the 1880s.

Erie — The idea of a 560 kilometers canal linking the Great Lakes with the Hudson River near Albany, NY, was considered "madness" by no less a man than Thomas Jefferson as early as 1809. Indeed, the task was formidable, with little precedent in the U.S. to dissuade such doubt. Prior to the Erie, the most ambitious canal projects in America were the Santee Canal in South Carolina and the Middlesex in Massachusetts. These canals were only 35 and 43 kilometers long, respectively. Considering their length and slow rate of construction (3 kilometers per year), the Erie, with its 585 kilometers length, must have seemed totally out of the question. Nonetheless, by 1816 New York began work on the project.

Several canals had been built in France and England by this time. The great French military engineer, Vauban, considered them important as a strategic transportation infrastructure useful for moving men and material during war time. In England, the canals brought coal to the factories of the Industrial Revolution. They moved the manufactured goods as well. The significant difference between Europe and the U.S. lay in the availability of trained professional engineers. Europe had several, America had few at best and none of native origin that could claim to have the knowledge or skill to build a canal such as the Erie. However, by its completion, America had its first generation of native-born engineers in the likes of James Geddes, Benjamin Wright, David Bates, Nathan Roberts, John Jervis and Canvass White. These men would build the canal by 1825 and in the process, invent the American civil engineer.

The Erie Canal route ran from Lake Erie at Buffalo to Albany (Figure 11-3). The canal was 40 feet wide and 4 feet deep through its 150 kilometers middle section from Utica to the Seneca River. Between Buffalo and the Seneca, the 250 kilometers western portion, were 129 kilometers east of Buffalo, where two significant engineering feats were accomplished. The first was the Irondequoit Embankment and the second, the Genesee River Aqueduct.

With the Irondequoit Embankment, James Geddes joined three natural ridges with two artificial ones, 402 and 70 meters, respectively. This allowed the canal to travel almost a mile (1500 meters) along its crest. The Genesee River Aqueduct was a stone construction with a span of 244 meters. The last impresssive feat of the western part of the Erie was the five double locks built at Lockport, raising the canal 18 meters. The work, completed in 1825, took 3 years.

Less impressive in an engineering sense, but daunting nonetheless, was the eastern portion, which fell 127 meters over the 175 kilometers from Utica to the Hudson River. Canvass White, the discoverer of a natural hydraulic cement to be used in the Erie's locks, showed why he was considered the finest of all the engineers to build the canal. By ingenious placement of locks, he controlled the route level and completed the work in just 3 years.[6]

Panama — The French began the canal in 1882 and worked for 20 years, spending $200 million and completing only 40%. The U.S. took over in 1903 for $40 million and completed the canal in 1920. It saved 40% of travel time from New York to San Francisco.

Major difficulties of the Panama Canal project were:

1. Medical — yellow fever and climate
2. Engineering — Chagres River (Atlantic side); Culebra Cut at Continental Divide (83 meters depth); six sets of locks, 34 × 300 meters, on each side of the Divide

Vasco Nunez de Balboa first proposed the construction of a canal linking the Atlantic and Pacific Oceans in 1513. Centuries passed in which economic and technical difficulties prevented such a massive undertaking. While many surveys were conducted through the years to establish possible routes, no construction was attempted until the late 19th century.

In March 1881 Ferdinand de Lessups, the builder of the Suez Canal, established the French Canal Company in an attempt to construct a sea-level canal across the Isthmus of

FIGURE 11-3.   Erie Canal at Little Falls (Library of Congress).

Panama. Work was halted in late 1888, with construction one fifth complete, due to lack of money. The failure of the French Canal Company was due to a combination of poor construction management and lack of good technical knowledge.[7]

The construction of the Panama Canal was taken over by the U.S. in early 1904. By 1924 the canal was opened to traffic. The U.S. engineers succeeded where the French had failed with a formula of better management, inventiveness, superior financing, and the use of large-scale sanitation.

The Panama Canal was constructed under the direction of the Panamanian Canal Commission from May 4, 1904 to 1915. During the planning and building of the project, problems were encountered with sanitary and municipal engineering, construction management, excavation and dredging, and with lock, dam and spillway construction. Solutions to these problems required new ideas and methods, and engineers to execute them.

From the Atlantic Ocean to the Pacific Ocean a vessel first passes through a man-made breakwater into Limon Bay. From Limon Bay it is lifted 26 meters above sea level by a series of three locks at Gatun Dam. It then proceeds for 39 kilometers across a man-made lake (Gatun) and thence 8 miles through Gailland Cut. Finally, the vessel then descends the Pedro Miguel and Miraflores locks down to sea level and the Bay of Panama.

Sanitation measures were needed to combat the diseases that stopped the French effort with 16,500 deaths.[8] Yellow fever and malaria, were responsible for over 16,500 deaths during construction by the French Canal Company. The U.S. realized the control of these diseases was essential to the success of the project. Col. W. C. Gorgas, a respected authority on tropical diseases, led the fight for the control and prevention of disease in the Canal Zone.

Yellow fever was a problem near the close of 1904, as it was in cities like Savannah. An outbreak of the disease caused panic among the workers. Control of the disease required eradication of the *Aedes* mosquito (the insect vector of yellow fever) by: (1) eliminating breeding places; (2) destroying infected mosquitoes; and (3) preventing mosquitoes from becoming infected by screening the patient in the infectious stages. Control was accomplished by: (1) providing an abundant supply of piped water (thus doing away with the need for open water containers used in collection of drinking water); (2) paving streets to eliminate puddles; (3) fumigation; and (4) screening of infected patients. With these measures, yellow fever was virtually eliminated in the Canal Zone.

Malaria was somewhat more difficult to control. The destruction of larvae by oiling and larvacidal poisons, protecting man from mosquito bites by screening and immunization, and the use of quinine were necessary. Steps which played an important role in the successful control of malaria by the engineers included elimination of breeding places through the filling of lowlands, construction of drainage ditches, the cleaning of stream and pond banks, and the cutting of brush. The significance of these preventive medicine efforts was a dramatic increase in the efficiency of the workers obtained at a small cost.

For the sanitation program of Col. Gorgas to work, an extensive municipal engineering program was developed. This program included the construction of roads, sewers and water treatment and distribution facilities. To allow for the drainage of water, roads were constructed with hard surfaces, crowned and graded (MacAdam). This most popular form of road construction on the Isthmus consisted of placing a 6- to 9-inch layer of crushed stone over a prepared subgrade, a thin layer of screenings was placed on top, followed by compaction with steam rollers. Drainage was provided by concrete curbs or graded ditches. Dust, a major problem of the macadam roads during the dry season (January–May) was controlled by application of water or oil to the road surface. In contrast, brick and concrete pavements were used in Panama City and Colon.[9]

House drainage and storm water runoff was carried in sewers. Both combined and separate sewage-water supply systems were built, using gravity flow and pumping to move the sewage to outfalls. Panama City used gravity flow for house and street drainage and discharged through pipes into the Gulf of Panama. Colon required a separate system for house and street drainage with sewage pumps, as elevations were insufficient for a gravity flow system.

Abundant supplies of clean, fresh water resulted from the construction of numerous water treatment and distribution facilities. Water purification plants were the most impressive feature of the water supply network. The treatment process, common today, consisted of aerating the incoming water, chemical treatment (alum addition) to promote settling, sedimentation, filtration, and distribution. Water was pumped by pipe throughout cities and towns. Amounting to less than 3% of the total cost ($13 million), the roads, sewers, and water supply systems played a major role in making the project possible.

DRY EXCAVATION AND DREDGING TECHNIQUES

The major dry excavation of the Panama Canal was the Culebra Cut. Passing through central Panama, crossing the Continental Divide, and with a length of 14 kilometers and a minimum bottom width of 90 meters, the excavation was accomplished by the use of explosives, steam shovels and trains. Earth slides and drainage proved to be other major problems during the excavations.

Explosives first loosened the earth and rock for removal by steam shovel. An accidental explosion occurred involving 18,000 kilograms of dynamite and resulting in 23 deaths. Other disasters were prevented by requiring the loading and firing of explosives on the same day as well as by careful handling and rotation of the dynamite. At the height of construction of Culebra in 1909, 68 steam shovels were in use. The shovels were used for loading loose earth and rock onto railroad cars. Excavation progressed through a series of step-wise cuts down to a final grade (Figure 11-4).

Earth from the excavation was loaded onto rail cars for dumping at other construction sites. Two types—open flat cars, emptied by an unloading plow, and automatic air dumping cars—were used. Rail lines were constructed along the length of the Cut, following the paths of the steam shovels.

The solution to the drainage problems required: (1) keeping the water out of surrounding areas, and (2) ridding the excavation of collected water. Channels were constructed on both sides of the Cut. Water from the area flowed into the channels and was carried to a reservoir or river away from the Cut. Water in the excavation area was carried off by gravity flow. Later, depths used in the excavation required pumps. The excavation always moved from low elevation to high elevation, allowing gravity flow to clear the excavation.

Slides were always a danger due to the instability of the geologic materials on oversteep slopes. Terracing the slopes on either side of the Cut aided in control of slides. Dredging was used in clearing channels through the canal (Figure 11-5). The dredging operations were divided into three divisions, the Atlantic, Pacific and Central.

The construction of the canal gave rise to bucket-conveyor dredges with bucket volumes up to 1.5 cubic meters.[10] These dredges provided fill and sand for the construction of Gatun dam, locks and the Colon breakwater. The Atlantic division dredges provided nearly 611,000 cubic meters of sand for the Gatun locks. The Pacific division concentrated on the excavation of the 152 meters channel through Miraflores Lake. Central division dredges dealt with the slides that troubled excavation of the Culebra Cut.

The Atlantic and Pacific divisions were charged with the construction of locks, dams and spillways. The Gatun locks had three locks for a lift of 26 meters. Gatun dam was an embankment dam similar to the British ones discussed earlier. Its difference lay in its great length—2500 meters; width—610 meters; and height—30 meters. At its completion, the Panama Canal was a significant engineering feat, surpassing the achievements of both the earlier Erie and Suez projects. It remains one of the most strategic canals in the world, along with the Suez. It was the result of the innovative use of sanitary and hydraulic engineering

FIGURE 11-4. Panama Canal during excavation of the Culebra Cut (Library of Congress).

FIGURE 11-5.   Dredge, Panama Canal (Library of Congress).

that began earlier with men like Chadwick, Leather, White, and the other great engineers of the 19th century.

AIR POLLUTION

Air pollution is not unique to our modern day civilization. In the Rome of 61 A.D., the Roman philosopher, Seneca observed:

> As soon as I had gotten out of the heavy air of Rome, and from the stink of the smokey chimneys therof, which, being stirred, poured forth whatever vapors and soot they held enclosed in them, I felt an alternation of my disposition.

In 1661, an author contemporaneous with Charles II wrote:

> The foul smoke which sullied all London's glories, superinduces a sooty crust upon all things that it lights upon, spoiling the buildings, tarnishing the plate gildings and furniture, and corroding the iron bars and hardest stones with those piercing and acrimonious spirits which accompany its sulphur and exciting more harm in one year than the pure air in the country could effect in hundreds—besides killing all our bees and making our gardens barren.

Sir Hugh Beaver[11] notes that in 1819 Parliament set up committees to "consider how far persons using steam engines and furnaces could work them in a manner less prejudicial to public health and comfort."

Jurisdiction in air pollution was placed in the hands of health agencies by the Public Health Acts of 1866 and 1875. It was not until the latter part of the 19th century that interest was generated in the U.S. In 1913 the "Commonwealth Club of California" sponsored a seminar on Smoke Problems in California. It is surprising how much was known at that time on the effects of air pollution.

F.G. Cottrell, in the seminars *Transactions*, clearly states the question of control of pollution in relation to sulfuric acid production from fume litigation at a copper smelter. Pointing out that after "driven to this, much against its will...it is making more off its acid rain than its copper output."

G.J. Pierce's paper dealt with the effects of acid rain (SO_2) on vegetation. Edgar R. Bryant dealt with smoke problems in large cities and listed control measures. Bryant enumerated the effects of carbon monoxide, hydrogen sulfide, and nitrogen oxides, particular nitrogen tetroxide (N_2O_4). Remember this was 1913! Professor Cottrell went on to develop and successfully market a commercial electrostatic precipitator. Although much was known about airborne contaminants in 1913, very little action was taken.

THE CASE OF DONORA, PENNSYLVANIA

1948 was a watershed year for air pollution management in this country. Due to the effects of smog in a small Pennsylvania mill town called Donora, the scope of air pollution management was raised from the local to national level. It occurred on a Friday night in October where participants and spectators complained that a low-hanging fog effected the outcome of a high school football game. By the following morning, 1440 people out of a population of 14,000 had sought medical care. By the end of the weekend 26 people were dead. By Sunday the smog cleared. The federal government was asked to investigate the disaster. The investigators were primarily industrial hygienists. They reported that 43% of population of Donora was affected that weekend, 14% severely. The most susceptible age group was 52 to 84 years. The investigators failed to find anything in the mills and the atmosphere. That shook them up. The problem was there were no bench marks for low-level, 24 h contaminations. Critical levels has been set for a 40 h week for strong, middle-aged workers. The need to understand the interaction of chemicals polluting the atmosphere

Table 11-1
Chronologue of Important Events in 19th and 20th Century Sanitary and Environmental Engineering

1829	First slow sand filters—London
1842	Edwin Chadwick—appointed Secretary Poor Law Commission-poverty and disease. ''Report on Sanitary Condition of the Labouring Population of Great Britain''
1843	Royal Commission on Health of Town appointed, Water supply and sewage
1855	John Snow—Broad Streat pump epidemic. ''On the Mode of Communication of Cholera.'' London
1855	First sewer system in U.S., Chicago
1843–1910	Robert Koch—discovered tubercle bacillus, tuberculin
1880	First seperate sewer system, sanitary and storm, Pullman, IL and Memphis, TN
1881	Carlos Finlay—advanced theory that mosquito transmits yellow fever
1888	Ludwig Wilhelm Winkler—dissolved oxygen test
1890	Louis Pasteur—fermentation, silk worm disease, anthrax, chicken cholera, rabies
1891	Sludge drying beds
1892	Hamburg cholera epidemic; proved the association between polluted water and disease
1892	Rapid sand filtration, Weston and Fuller
1900	Major Walter Reed—yellow fever studies, Cuba
1905	First edition of *Standard Methods*
1908	Chlorination of water supplies
1914	Activated sludge process introduced by Arden and Lockett
1917	First activated sludge plants in Withington, G.B. and Houston, TX
1948	Donora, PA air pollution disaster
1951	Biosorption process of activated sludge
1952	London killer fog, 4000 deaths
1955	U.S. Air Pollution Control Act
1957	Genetic hazard of radiation incorporated into standards
1963	U.S. Clean Air Act
1965	U.S. Motor Vehicle Air Pollution Act
1967	U.S. Air Quality Act
1970	U.S. National Environmental Policy Act; Funding of U.S. Environmental Protection Agency (EPA)
1972	U.S. Clean Water Act
1972	U.S. Marine Protection, Research and Sanctuaries Act
1976	U.S. Resource Conservation and Recovery Act

and their effects on people, plants and animals, underlined by the disaster at Donora, prompted the new technology of air pollution control.

Further excellent work in industrial hygiene and air pollution was carried out by the U.S. Public Health Service, with some of the finest epidemiological studies ever made being done by this Service between 1925 and 1935. These studies included long-term manganese poisoning, lead battery manufacture, carbon monoxide in the Holland Tunnel and the Lead-arsenic study in Washington apple orchards.

The effect of toxic substances on fish were first conducted by Penny and Adams in 1867.[12] Shelford and Wells in 1912 did the first work in the U.S. E.B. Powers' classic paper, ''The Goldfish as a Test Animal'' in *The Study of Toxicity* in 1917.

While these early developments in the scientific and technical study of environmental problems are important to the development of modern environmental science and engineering (see Table 11-1), it remained the province of reformers like Chadwick of England, von Pettenkofer of Bavaria, and Florence Nightingale to revolutionize sanitation and public health.[13] Their reaction to filth, dirt, crowding and deleterious conditions is such, in Dubos words, ''No medical discovery made during recent decades can compare with the introduction of social and economic decency in the life of the average man''.[14] Engineering has played a key role in this endeavor.

NOTES

1. **Snow, J.** A reprint of two papers by John Snow, M. D. (The Commonwealth Fund, New York, 1855.)
2. **Hazen, A.** *Clean Water and How to Get It,* 2nd ed. John Wiley & Sons (1907).
3. A major dam is defined as above 15 meters in height or between 10 and 15 meters with a crest length of 500 meters, impounds a million cubic meters, flood discharge not less than 2000 m, difficult foundation problems, or is of an annual design (Register Criteria for International Commission Large Dams).
4. *Los Angeles Times,* March 19, 1978.
5. **Sandström, G. E.** *Man the Builder.* McGraw-Hill (1970), pp. 208–211.
6. **Tarkov, J.** Engineering the Erie Canal. *American Heritage of Invention and Technology,* Vol. 2, No. 1 (1986), pp. 50–57
7. **Martinez, O.** *Panama Canal,* Gordon & Cremonesi, New York (1978).
8. **Mason, C. F.** Sanitation in the Panama Canal Zone. *The Panama Canal,* McGraw-Hill, New York (1916).
9. **Wells, G. M.** Municipal engineering and domestic water supply in the canal zone. *The Panama Canal.* McGraw-Hill, New York (1916).
10. **Straub, H.** *A History of Civil Engineering,* M.I.T. Press, Cambridge, MA (1964).
11. **Beaver, H. E. C.** The growth of public opinion. *Problems and Control of Air Pollution.* Mallette, F. S., Ed. Reinhold (1955).
12. Fourth Report of the Royal Commission of Pollution of Rivers in Scotland. (1867).
13. **Shiffman, M. A.** Food and environmental sanitation. In: *History of Environmental Sciences and Engineering.* School of Public Health. University of North Carolina at Chapel Hill. Department of Env. Sci. and Eng. Pub. No. 259 (1971), p. 41.
14. Ibid., p. 41.

12 20TH CENTURY ENGINEERING: PART 1

The 20th century has given birth to those disparate fields of engineering beyond those already in place at the beginning of the century. Of American Engineering Societies these were:

1. American Society of Civil Engineers, 1852
2. American Institute of Mining and Metallurgical Engineers, 1871
3. American Society of Mechanical Engineers, 1880
4. American Institute of Electrical Engineers, 1884
5. American Institute of Chemical Engineers, 1908

This list has grown with the evolution of petroleum engineering, aerospace engineering, transportation engineering, ocean engineering, nuclear engineering, environmental engineering, etc. (Table 12-1). In the beginning of the century the most profound innovation was the mastery of powered, manned flight (1903). This was made possible by the industry of Orville and Wilbur Wright but the fuel that powered their home-made engine was made possible by James H. Young's development of petroleum refining by distillation in 1847.

In the development of skyscrapers we see the final conquest of steel over stone. The first iron frame factory (1779) was British but the ultimate expression of this potential in building was the result of American structural engineers and architects in the early 20th century. In environmental and chemical engineering the treatment of wastewater, hazardous wastes, distillation of petroleum and polymer science have grown since the 1920s.

WORLD WAR I AND TECHNOLOGY: AIRPLANES—POWERED GLIDERS TO WARBIRDS

The design of the airplane of Orville (1871–1948) and Wilbur (1867–1912) Wright was hardly the standard for successful commercial use of the invention. Based on Octave Chanute's (a French-born American civil engineer, 1832–1910) extensive glider experiments,[1] the first powered biplane consisted of a framework of wood, stretched canvas, and a home-made four-cylinder, 12 horsepower motor (81 kg). The total weight of the FLYER was 340 kg. The two rear-mounted chain-driven propellers turned in opposite directions about 3 meters. The wing span was 18.1 meters with a total wing area of 47.4 sq. meters. The wing chamber was 1/20. It had both vertical rear rudders and front elevator. Top speed was approximately 48 kph. The undercarriage was a set of skids. The FLYER made four flights on December 17, 1903, with distances of 37, 53, 61, and 260 meters. The fourth flight lasted 59 seconds (this latter figure implies an in-flight distance of 791 meters at 48 kph). The Wrights evolved a successful, practical flying machine in 1904–05—the FLYER, No. 3. Forty-nine flights were made in 1905 with one for 38 min 3 s covering 39 km at 61 kph.

Europe failed to match the success of the Wrights, due to bad copying of their design. This lack of success had far-reaching effects on aviation history. Gibbs-Smith (1960) gives five reasons why the European could not duplicate the Wrights' success (the French engineer, Esnault-Pelterie, in 1905 claimed he had proved their claims false). These were:

1. There was insufficient data for them to duplicate the Wrights' glider (1902).
2. They did not understand how it operated or the aerodynamic reasons for its features.

TABLE 12-1
American Engineering Societies

Society names	Founding date
American Society of Civil Engineers (ACSE)	1852
American Institution of Architects (AIA)	1857
American Institute of Mining and Metallurgical Engineers (AIME)	1871
Society of Mining Engineers of AIME (SME)	1871
American Society of Mechanical Engineers (ASME)	1880
Institute of Electrical and Electronic Engineers (IEEE)	1884
American Society of Naval Engineers (ASNE)	1888
Society of Naval Architects and Marine Engineers (SNAME)	1893
American Society of Engineering Education (ASEE)	1893
American Society of Heating, Refrigeration and Air Conditioning Engineers (ASHRAE)	1894
American Society of Testing and Materials (ASTM)	1898
Society of Automotive Engineers (SAE)	1905
American Society of Sanitary Engineers (ASSE)	1906
Luminating Engineers Society of North American (LESNA)	1906
American Society of Agricultural Engineers (ASAE)	1907
American Institute of Chemical Engineers (AICE)	1908
American Society of Safety Engineers (ASSE)	1911
Society of Petroleum Engineers (SPE)	1913
American Society of Metals (ASM)	1920
American Institute of Aeronautics and Astronautics (AIAA)	1932
Society of Manufacturing Engineers (SME)	1932
National Society of Professional Engineers (NSPE)	1934
American Society of Quality Control (ASQC)	1946
American Nuclear Society (ANS)	1954

3. They did not possess the thorough theoretical and practical grounding in aeronautical problems which the Wrights had gained by extensive labor and experimentation since 1899.
4. They had no extensive glider experience.
5. The Europeans were impatient and underestimated the difficulties of flying and simplified the problems involved. This impatience kept them from working from first principles.

From 1905 to 1908, the Europeans wasted a prodigious amount of effort. In 1907 the two forms of the early airplanes were clearly established—the pusher biplane and the tractor monoplane. The latter form was championed by Louis Blériot (1872–1936). Henri Farman (1877–1964) also pursued the monoplane design. On August 8, 1908, Wilbur Wright used No. 7 FLYER to totally confound and disprove their distractors when he flew it at Auvours, France. The year 1909 saw the airplane come of age. The outstanding event of that year was Blériot's flight of the English Channel on July 25. He flew his No. IX monoplane with a three-cylinder 25 hp motor. He flew the 37.8 kilometers in 36.5 min. His flight had a profound effect on governments in Europe who saw the role of the airplane in warfare. 1911–1912 marked the first large-scale and official concern with the airplane's role in warfare; several important technical advances, including monocoque construction, the first flying-boat, the first flight at over 100 mph, and the extended use of metal in aircraft structures.

The tractor biplane was introduced in Britain, the AVRO biplane with the designs of F.M. Green and the great Geoffrey de Havilland (1882–1965), whose company was to build the first commercial jet liner, COMET, in the decades following World War II. The first use in warfare was 1911 with a reconnaissance flight, by Italian Captain Piazza in a Bleriot, over Turkish positions near Azizia (Tunisia). In 1912 Geoffery de Havilland designed the B.S. I, the first fast single-seater in history, and ancestor of every fighter aircraft thereafter. Its most direct progeny was the S.E. 5.

Monocoque construction of the fuselage was a major structural innovation in which the skin (or shell) carries all or most of the loads, as opposed to a framework fuselage left open or covered with wood or fabric material. The first wood monocoque structure was designed in France by Béchereau. H. Reissner introduced corrugated alumimum wings in his canard-type monoplane. One of the Curtiss flying-boats was the first airplane to be catapult-launched, from an anchored barge with a compressed air launcher. A Royal Flying Corps was founded in Britain (1912). France and Germany followed suit and France experimented with bombing gear. A U.S. officer, Captain Albert Berry, made the first parachute jump from an airplane (1912). A Morane-Sauliner climbed to 5610 meters.

By the year 1914:

- Lawrence Sperry demonstrated first gyroscopic stabilizer.
- Sperry later invented the turn and bank indicator.
- Planes were fitted with radio.
- Fighter-reconnaissance aircraft-type and bomber-type evolved quickly from the simple observation variety.
- Anthony Fokker (1890–1939) revolutionized air warfare with fixed forward-firing machine gun shooting between the propeller blades by means of an interrupter gear used on 'EINDECKER' fighter.
- France countered the Fokker design with the NIEUPORT in 1915.

By 1918 the best fighters on each side (S.E. 5a, Fokker D. VII, Nieuport 28) had a service ceiling of 7620 meters; top speed of 250 kph (225 kph at 3000 meters) with a rate of climb of 300 meters/min (1000 ft/min) with a duration of 3 h flight. Bombers had a range of 2091 kilometers with a bomb load of 3400 kg. Aerial photography had made great strides and instrumentation likewise. Bombardiers used corrected bomb sights. Pilots had anti-spin controls. Aircraft were fitted with superchargers, variable-pitch propellers, rockets and incendiary weapons, radio-controlled aerial bombs, and radial engines. Hugo Junkers (1859–1935) of Germany was to influence the whole course of aircraft evolution with these advances in design: cantilever wing, first all-metal monoplane, first low-wing monoplane, the Junkers D-1.

Aircraft performance of this era can be quantitatively compared by use of the following equation

$$H = \sqrt{\frac{P_o}{f} \left(\frac{\zeta}{\sigma}\right)}$$

where P_o is the maximum power at sea level, f is the drag area, σ is the atmospheric density ratio for a given altitude, and ζ is the percentage of maximum power at that altitude.[2]

After the war these leading designers moved into commercial aviation. The Farmon brothers initiated the first commercial passenger service between London and Paris in 1919 with converted GOLIATH bombers. de Havilland produced the first commercial passenger aircraft designed for the purpose in 1919 as well, the D.H. 4a transport. Junkers and Fokker developed three-engine monoplane designs for commercial use with a Fokker tri-motor design, the F.VIII B making the first trans-U.S. flight in the early 1920s.

This design was eclipsed by the first modern commercially successful airliner, the DC 1, which quickly evolved into the DC 3. This aircraft was an accumulation of prior innovations—all metal, monocoque, retractable landing gear, swept-back single wing—which together dictated design from 1936 to 1950.[3, 4] Of these innovations, Jack Northey's wing was the most singular engineering design. Its swept-back shape, and web-like internal structure allowed it to carry the weight of the plane in flight with its upper surface in tension.

WORLD WAR I AND TECHNOLOGY: THE SUBMARINE—EARLY HISTORY

The idea of the submarine was not new. A crude one-man submarine (TURTLE)(Figure 12-1) was invented by an American engineer named David Bushnell (1742–1824) and used unsuccessfully in the Revolutionary War in New York Harbor. Robert Fulton (1765–1815), of CLERMONT fame, was an exponent of submarines in the early 1800s. In 1800 Fulton built NAUTILUS which dove to a depth of 7.6 meters for 17 min in the Seine River. Sent to England in 1786 by Benjamin Franklin, ostensibly to sharpen his skills as a painter, he turned his back on art and took up engineering.

The NAUTILUS was 7 meters with a 1.9 meter beam. She was shaped like an ellipsoid. She had a three man crew who used a hand crank to move a single propeller. She would submerge by flooding her hollow iron keel. Angle of dive was controlled by horizontal fins on the rudder. Pumping the water ballast from the keel allowed her to surface. In later experiments in the ocean off LeHavre (France), she stayed submerged for 6 h. The NAUTILUS first had an archimedean screw, which Fulton changed to a four-bladed propeller.

Fulton's work on submarines ended in France in 1804 and later in England in 1806. Frustrated, he returned to America where he reached success with the design of the steamboat CLERMONT. Before he left England he deposited all his notes with the American consul. In 1920, William B. Parsons, a Fulton biographer, published these notes. Fulton's final British design was a 10.6 meter boat with a 3 meter beam carrying a six man crew and provisions of 20 days at sea. It was armed with 30 mines. It had a mast for surface running and a two-bladed propeller which could be folded up out of the water, during surface operation, to reduce drag. It had ventilation pipes and a conning tower with view ports.

The French commission's report on Fulton's "diving machine" was summed up thusly: "The underwater boat...is a means of terrible destruction because it operates in secret in an almost unavoidable manner." World War I was to prove the accuracy of this conclusion. During the American Civil War, a Confederate design, also propelled by man-power, sank the Union warship, HOUSATONIC. Unfortunately for the crew of the HUNLEY, the Confederate submarine, it was also a casualty of its own successful attack.

In the 1870s, the Whitehead locomotive torpedo developed from an Austrian design (1864) by British engineer John Whitehead (1823–1905), became available. It allowed the attack of surface ships without committing suicide like the HUNLEY. In addition to the development of the torpedo, electric batteries and motors were becoming more reliable and offered the submarine designs 50 times the mobility of a hand-powered submersible. France, throughout the 1880s, pursued submarine development as an official naval policy to meet strategic needs. Boats of this period, in the U.S., used a combined steam-battery propulsion system developed by John P. Holland (1840–1914).

Holland's first successful vessel was the FENIAN RAM, built in 1881, cigar-shaped, 10 meters long, with a 1.8 meter beam. Her hull was 0.2 meter flange iron. Holland powered her with a 15-17 hp, two-cycle Brayton gasoline engine located amidships. Sealed bow and stern compartments contained compressed air for positive buoyancy. A relatively useless pneumatic gun, 3.3 meters long, was the submarine's armament. Her clean hydrodynamic shape foreshadowed modern designs.

Germany did not build a submarine until 1906. Two reasons are given, historically for this late entry:

1.   Admiral Von Tirpitz emphasized the development of a surface battle fleet to rival Britain.
2.   The German Navy objected to the use of steam or gasoline for propulsion.

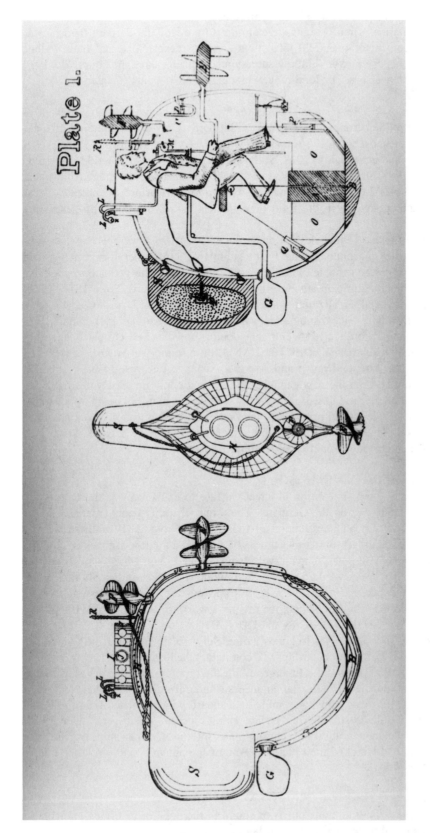

FIGURE 12-1. Bushnell's TURTLE (U.S. Navy archives).

With the development of the Korting heavy-oil (diesel) engine the Navy had no technical objection to the submarine. Their designs were similar to the Holland-derived British D.1. series with a displacement of over 500 tons, twin screws, diesels, external main ballast tanks, and extended battery power to allow submergence during daylight hours. By 1914 the Germans had 29 "overseas" or long-range vessels, called U-boats ("untersee boots") ready for World War I.

The success of the U.S. Navy's anti-submarine strategy and convoy system allowed for the transport of two million troops to Europe from the U.S. Only a few hundred of these men were lost at sea, the most being 175 Americans when the Cunard steamship TUSCANIA was torpedoed off Ireland in February, 1918. Still, throughout the years 1914–1916, German U-boats cruised at will off the English coast. In 1915 the "exchange rate", i.e., ships sunk to U-boats destroyed was 23:1. The other modern innovation to warfare, the airplane, was then used against the submarine in the form of seaplanes and flying boat anti-submarine patrols.

In 1916 the hydrophone (under development since 1915) was developed against submarines. This underwater microphone could hear a submarine's machinery or propellers and was the first electronic device capable of detecting a completely submerged submarine. The depth charge was another significant advance in anti-submarine warfare. From 1916–1918, with German High Command's realization that they could not break the British Grand Fleet's superiority, they opened unrestricted submarine warfare. Casualties in Allied merchant shipping rose to over 115 ships/month in 1916. Germany demonstrated its new U-boat's capability by sending a commercial submarine, DEUTCHLAND, on a round-trip voyage to the U.S. Using U-boats with this extended range and size the exchange rate rose to 65:1 in 1917. The submarine did not become the new capital ship in that it could not easily destroy other warships. The importance of the submarine as a naval weapon lay in its not being invulnerable but rather being replaceable. It did not replace the battleship as the arbiter of sea power as its protagonists hoped. That would wait for the nuclear age, after the second World War and the coming of another new vessel type, the aircraft carrier.

WWI AND TECHNOLOGY—"DAS BOOT"

In 1914 the most singular innovation in naval warfare to that time was the large scale offensive use of the submarine by the Germans. The threat of the U-boats created a whole technology for the detection and destruction of these vessels, but in 1914 there was no method for detecting submerged submarines unless they showed their periscopes or conning towers. Submarine armament was comparatively weak as well. Torpedoes were fired singularly and were of short range of 0.75 meter. Shallow-draft vessels and small ships were hard to hit due to the depth that most of the early torpedoes ran. Nevertheless, the U-21, in September 1914, became the first submarine to sink a warship, H.M.S. PATHFINDER, since the sinking of the HOUSATONIC by the HUNLEY.

The impact of the submarine was to almost completely isolate Britain from crucial sea resupply, particularly by U.S. merchant ships. To counteract the loss of American transports, the U.S. Shipping Board organized the Emergency Fleet Corporation and every shipyard in the country was put at top speed. Industrial engineers standardized designs and inland steel plants turned out parts for these ships. Assembly at seaboard points reached six ships being launched per day. Specialized sub-chasing surface ships such as "destroyers" were built, emphasizing speed and maneuverability over armor plate. Interned German ships were seized. Although their crews had wrecked their engines, American engineers and workmen reconditioned them successfully.

SONAR

Active sonar is a device that emits bursts of sound in the hope of receiving an echoing answer. This acoustical device can easily give both range and bearing of targets within its range.[5] Echo-ranging sonar was developed a decade or more before World War II. This system utilized a relatively narrow beam (15–20°). Coupled with a speed of sound in water of 914 meters/s, the narrow beam transducer required accurate training of the projector and hence was somewhat time consuming in usage. This disadvantage was offset by the fact that electromagnetic waves do not propagate in sea water; hence, sonar is the only reliable means of searching for submerged submarines or objects in general.

Pulse length has an important bearing on sonar efficiency—the shorter the pulse (10 ms), the more accurate the range information on the target. Conversely, the longer the pulse (100 ms), the greater the detection capability. Shape and intensity of a pulse reflected from a target is dependent upon the aspect and shape of the target. For instance, if a submarine presents only its bow to the impinging pulse the sonar echo is minimal. The echo from the quarter is stronger, with the stern less but greater than the bow due to propeller noise.

Most early sonar transducers were piezoelectric devices where changes in a crystal's physical dimensions are induced by electrical voltage applied across it. Likewise, a returning pulse stresses the crystal, generating a weak voltage due to compression. The same electrical connections that apply voltage can be used to indicate the existence of a received pulse. Typical beam patterns are characterized by widths in "dB" or decibels. Thus, a 20 dB point refers to a logarithmic characterization of beam spread or the power radiated along that angle is 100 times less than the amount transmitted directly along the "beam". A 10 dB spread is thus 10 times less radiated power than that along the beam.

WORLD WAR II AND TECHNOLOGY—THE SUBMARINE CONTINUED

During the decades following World War I, designers increased the submarine's range, speed, and depth. Attack-warning and detection systems were added and improved. The submarine's one weakness, often fatal, was the need to come to the surface to recharge batteries. By 1942 Allied patrol aircraft had radar. This meant that a submarine was vulnerable to attack at night and in inclement weather. To circumvent this, German designers developed two separate approaches: (1) The "Schnorchel" and (2) the Walther-cycle engine.

The "Schnorchel" was a pair of fixed air pipes, one for air intake and the other for exhaust (Figure 12-2). It allowed the submarine to run submerged but close to the surface but here, too, the housings for the tubes were detectable by very high frequency radar.

The Walther-cycle engine, developed by Dr. Hellmuth Walther, used hydrogen peroxide, broken down by catalysts into water and oxygen as fuel for the diesels. The diesel oil burned the oxygen at high temperature, which in turn turned the water, injected into the combustion chambers, into high pressure steam. The steam and burned-oil gas were used to power a turbine which propelled the vessel. The boat to use these engines was designated Type XXVI. None were completed before the war ended but British versions of the early 1950s had performance that exceeded conventionally powered vessels. The Walther-cycle engines may have been the standard propulsion for the world's submarines had it not been for the advent of the nuclear-powered submarine in 1954, the U.S.S. NAUTILUS (recall Fulton's first submarine). Following World War I, the British pushed forward on counter measures for submarine detection. ASDIC was the first real improvement over the simple hydrophone.

ASDIC was an acoustic system which, unlike the hydrophone, had an active transducer that propagated a supersonic "ping" that reflected off metal hulls and gave echoes at ranges of over a kilometer. The British put too much confidence in the locating ability of the ASDIC and did not push beyond this initial system. German U-boats simply attacked, at night, on the surface. Not until radar-equipped patrol airplanes were combined with ASDIC-equipped ships did anti-submarine measures become reasonably successful. U-boats, in 1942, sank $7^3/_4$ million tons of shipping, but this was offset by 7 million tons of new ships built by the allies. By May, 1943, Germany had lost 250 of 650 U-boats to radar, ASDIC and coordinated convoy defense methods. After 1943 to the end of the war, they lost another 534 boats. The U-boats sank a total of 2828 merchant vessels.[6] In the Pacific American submarines

FIGURE 12-2.   Type 21 U-Boat with schnorchel (U.S. Navy archives).

destroyed one third of the Japanese Navy. Prior to 1944 Japan had no radar in ships or aircraft. American submarines successfully destroyed Japanese commerce, in comparison to the failure of Germany's attempt to do the same with Britain. Advances in technology played the key role in both theaters.

Almost all advances, since World War I and II, in the enhanced performance of sonars have come in the field of transducer technology. New insights into the propagation of sound in the ocean has led to the demand of efficient low-frequency transducer elements of great power output to echo-range at greater distances. Many precepts of information theory have had a great impact on sonar.

The ideal shape for broadest array coverage is spherical. These type arrays are being deployed in modern nuclear submarines and can be used for both echo ranging and passive listening. Passive sonar devices can determine range and bearing without active pulse transmission, employing a frequency analysis of the sound received by Fourier Analysis[7] utilizing digital computers. The analysis samples the wave for magnitudes at various points along it using frequency vs. time to compute ranges. It has become indispensable to modern engineering analysis, ranging from offshore structural design to the thermodynamics of rocket motors.

RADAR

In the 1930s the Lorenz Company had developed a blind-approach (R.D.F.) system for aircraft. Basically, it was a transmitter that used two adjacent radio beams, one broadcasting Morse dots, the other Morse dashes. Where the two beams overlapped provided a steady note for guidance to the transmitter. Germany developed this system into the "X-Gerät" (X-device) for bomber guidance developed by Dr. Hans Plendl. Britain developed a similar system called "GEE". Telefunken, one of Germany's electronics firms, developed a system to compete with the Lorenz Company's (Plendl) "X-Gerät". This was the "Knickebein" system which used only two Lorenz beams rather than four by Plendl's device. It was less accurate but simpler to use by bombers.

Radar, the real technological advance in electronics in World War II, was based on a 1904 invention called a "Telemobiloscope" or a "hertzian-wave projecting and sending apparatus". The inventor was a German, Christian Hülsmeyer. This early device had a range of several hundred yards but could not determine directionality of returned waves or signals. Further, it had no way to amplify the returned, weak signal. Hülsmyer's invention was not significantly improved for 25 years. In 1933 Dr. Rudolph Kühnold, head of the German Navy's signals research department, was working on an early version of sonar. Kühnold intuitively reasoned that what worked underwater with sound waves might work above the surface with radio waves. This intuition is fundamentally correct, as a recent comparison of the two systems permits the use of similar quantitative equations.[8] The empirical analogy shown here allowed Kühnold to construct a continuous signal radar transmitter at Kiel with which he bounced signals off the battleship HESEN. Kühnold had reinvented Hülsmeyer's "Telemobiloscope". Kühnold later switched to pulsed transmissions, an important improvement. This allowed the time comparison in transmitted and received pulses for accurate distance (range) measures.

The German radar was called "D.T. Gerät", D.T. stood for Dezimeter Telegraphie (10 cm telegraph).[9] The 1936 version had a range of 48 kilometers (30 miles). The Gema unit's operating frequency was changed from 150 to 125 megacycles with a range of 80 kilometers (50 miles) and was known as "Freya" radar. Telefunken entered the radar field in 1936 and produced the "Würzburg" radar. It worked at 560 megacycles with higher resolution which anti-aircraft batteries would need. Britain's equivalent to Germany's "Freya" was "Chain Home" and could detect aircraft at 200 kilometers ("Freya" had a 120-kilometer range) and determine altitude ("Freya" could not). "Freya" radar had 360° coverage while "Chain Home" had only a fixed 120° arc.

British engineers developed the most significant advance in radar with the cavity magnetron. This device stabilized the radar's microwave frequency while amplifying it as well. Accurate 10 centimeter radar and the later high resolution 3 centimeter radar were the result. While the magnetron was a British invention, it was for American engineers working at the MIT-based "Rad Lab" (Radiation Laboratory) that advanced applied radar technology perfecting 10 centimeter technology and developing that of 3 centimeter.

An improved "Würzburg" was used in directing German night-fighter aircraft. Germany's (or Britain's for that matter) best radar could give range, height, and bearing up to 240 kilometer. The "Lichenstein" radar was aircraft mounted, with a range of 183 meters to 3.2 kilometers. British bombers utilized a system called I.F.F. ("identification, Friend-or-Foe") which was designed to receive pulses from British ground radar at 25 megacycles and reply with a coded pulse on the same frequency, preventing confusion on the part of British air defense. Later in the war, German radar units could trigger I.F.F. units and track British bomber formations. Britain's most significant airborne radar was "H$_2$S", the prototype of rotating-sweep antennae radars of today. It was one of the first radars to use the magnetron rather than the Klystron valve for frequency control. "Oboe" was the British airborn short-range guidance radar equivalent to "Lichenstein". German fighters were equipped with "H$_2$S" radar transmission and used the improved "Lichenstein SN-2" radar to attack the bombers. The "SN-2" radar was not effected by "Window" (aluminum foil) jamming used by the British. Another homing device used by Germany was "Kiel-Gerät", an infrared system that detected a bomber's engine exhaust from up to 6.4 kilometers (4 miles).

WORLD WAR II AND TECHNOLOGY—THE ROCKET GROWS UP: PEENEMÜNDE

Two important monographs appeared on rockets post-World War I; one was by an American, Robert H. Goddard (1882–1945) and the other by the German Hans Oberth, in 1923.[10,11] Goddard was an experimentalist, whereas Oberth was a theorist. Of the two, only Goddard successfully flew a liquid-fuel rocket (O$_2$ and gasoline). This took place in March, 1926. It flew 56 meters in 2.5 s. The decade of the 1930s saw little success (or even experimentation) on liquid-fueled rockets except by Goddard and German workers at the Raketeflugplatz ("Rocket port") in Berlin. These men included Wernher Von Braun (1912–1979), Willy Ley (1906–1969), Walter Dornberger (1895–1980), and Albert Peitsch.

Using Oberth's formula for rocket thrust:

$$P = c \cdot dm/dt$$

where p = thrust, c = exhaust velocity, and dm/dt = mass of exhaust/s.

The rocket men began work in 1932. By 1933, they had flown a rocket to about 2000 meters. In 1937 they transferred their work to Peenemünde on the Baltic coast. Two principal missile weapons were developed at Peenemünde—the V-1, a pulse-jet glider bomb and the A-4 (later V-2), a true rocket. The V-1 had a fuselage 7.7 meters long with a 1000 kg warhead. It flew on 80-octane gasoline, with a fuel tank capacity of 567 liters. Fuel consumption was one gallon per mile, with a speed of 580 kph. It had an automatic pilot mounted in the tail with controls driven by compressed air. Its wing span was 5.1 meters.

The pulse-jet was the invention of a German engineer named Paul Schmidt. The jet was a simple design consisting of a pipe or tube open at the rear and closed in front by spring-mounted shutters which were opened by exterior air pressure. In flight the air pressure forced the shutters open, activating a fine fuel spray which ignited, forcing the shutters closed again. The combustion gases rushed out the open rear of the pipe propelling the V-1. To launch the missile required a ramp with a piston powered by the decomposition hydrogen peroxide. The pulse jet took over after launch. When out of fuel, a clock mechanism, set

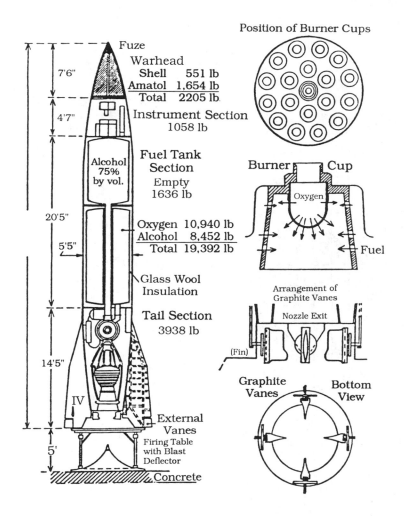

FIGURE 12-3. A-4 (V-2). *Left:* Cross-section of the whole rocket on its firing table. *Right, top to bottom:* Position of the eighteenth burner cups in the top of the motor; cross-section of a single burner cup; side view and bottom view of the exhaust nozzle, showing arrangement of the graphite vanes.

for this time, locked the control surface into a dive configuration whereupon the warhead exploded on impact.

The A-4 (V-2 or "Vengence Weapon-2") was built as a pure rocket with a range of 257 kilometers, carrying a 1000 kg warhead. The A-4 rocket consisted of four main sections: warhead, instrument compartment, fuel tank, and tail section with motor (see Figure 12-3). The empty weight was 3991 kilograms. The loaded rocket weighed 12836 kilograms and was 14 meters in length. The fuel, alcohol and liquid O_2, was pumped into the motor by a centrifugal pump driven by steam. This steam was generated by potassium permangenate ($KMnO_4$) and hydrogen peroxide (H_2O_2). The motor used film cooling (a film of "cold" alcohol) outside the burner cups to protect against a burn through of the motor. The rocket was fired in three stages: electrical ignition of a pinwheel ("Zündkreuz"), preliminary stage (gravity flow of fuel only), and main stage (pumped fuel with 27 tons of thrust).

Figure 12-4 shows the A-4 (V-2) performance as a long-range rocket. Its maximum range was just over 300 km. Here the altitude given is just under 100 km (Post-war firing of captured V-2s established a maximum altitude for the rocket at 180 kilometers).

Steering of the V-2 was accomplished by graphite vanes that projected into the exhaust

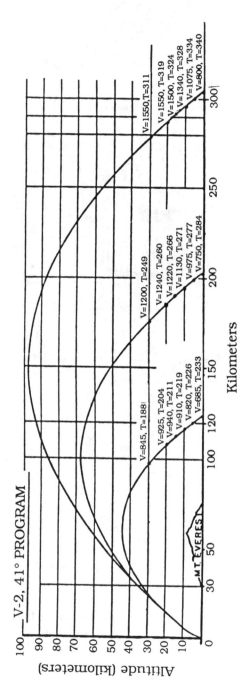

FIGURE 12-4.   Long-range trajectories.

TABLE 12-2
V-2 Firings, The Hague, 1944–45

Month	V-2s fired	Failures
September 1944	24	0
October 1944	81	4
November 1944	142	12
December 1944	132	17
January 1945	229	15
February 1945	207	12
March 1945	212	19
Total	**1027**	**79 (7.7%)**

blast. After "brennschluss" ("loss of burn"), steering was done by the stabilizer vanes in the air stream. The "burn" lasted about 1 min, with the speed of sound reached at 25 s. Maximum velocity was around 1500 m/s. The performance of the V-2 was remarkably reliable for such a relatively untried concept in warfare, i.e., the long-range ballistic rocket. The following excerpt gives tabulated data on the V-2s' wartime performance.

Two Dutch scientists, Professor Uytenbogaart and Dr. Kooy, tabulated the number of V-2s fired from The Hague and its suburbs and the failures which could be observed from the area of the firing site (Table 12-2).

The German V-2 offensive came to an end at 1645 h (4:45 p.m.) on March 27 (1945) with the 1115th rocket to fall in England. The campaign lasted 7 months. During that time the Germans had fired at least 1300 V-2 rockets at London and some 40 or more at Norwich.

The wartime development of the rocket reversed on the projected evolutionary trend by placing the emphasis on long-range capability over that of high altitude. Post-war development returned to an emphasis on the latter area. Working in Germany during the pre-war period, Ludwig Prandtl (1875–1953) was a leading theorist in what would become the field of aerodynamics. Prandtl has been called "the father of aerodynamics", due to his early work on the boundary layer on surfaces over which fluids flow. This discovery in 1904 led to a better understanding of friction and drag on airfoils. This led to streamlining of aircraft and, together with the Briton F.W. Lanchester, introduced Wing Theory.[12] Prandtl studied compressible and supersonic flow in wind tunnels, another first.

At the University of Göttingen during Prandtl's tenure was the Hungarian-born engineer Théodore von Kármán (1881–1963). His work on turbulent flow over various shapes led to Vortex Theory. It was Kármán who analyzed the failure of the Tacoma Narrows Bridge, called "Galloping Gertie", and determined oscillations produced by a 67 kph (40 mph) wind were excited by vortex shedding frequency.[13] Kármán worked on propeller design and the aerodynamics of gliders. His work laid foundations for the Luftwaffe of Nazi Germany of which he disapproved. In 1929 he joined Cal Tech and helped found the Jet Propulsion Laboratory (JPL).

WORLD WAR II TECHNOLOGY—THE JET

Flight by a heavier-than-air craft is possible by the combination of a structurally sound frame, a stable airfoil, control elements, and a power source such as an internal combustion engine. The objective of sound aeronautical engineering from the Wrights onward has been to achieve maximum strength in minimum structure with high efficiency in propulsion and guidance.

The propeller-driven (pulled) piston engine aircraft reached its apogee in the second World War. Designs flew higher and faster but clearly they were reaching a limit which

TABLE 12-3
Examples of Propeller-Powered Speed Records

1.   1913, Absolute speed record of 126.64 mph established by French Deperdussin landplane
2.   1920, Absolute speed record of 194.49 mph established by French Nieuport 29V landplane
3.   1923, Absolute speed record of 267.16 mph established by American Curtiss R2C-1 landplane
4.   1927, Absolute speed record of 297.83 mph established by Italian Marcchi M-52 seaplane
5.   1931, Absolute speed record of 406.94 mph established by British Supermarine S-6B seaplane
6.   1934, Absolute speed record of 440.60 mph established by Italian Marcchi-Castoldi MC-73 seaplane
      (this record for propeller-driven seaplanes still stands and is unlikely to be surpassed in the near future)
7.   1938, Absolute speed record of 469.22 mph established by German Messerschmitt 209V1 landplane
8.   1969, Absolute speed record of 483.04 mph established by highly modified American Grumman F8F
      landplane

From Loftin, 1985.

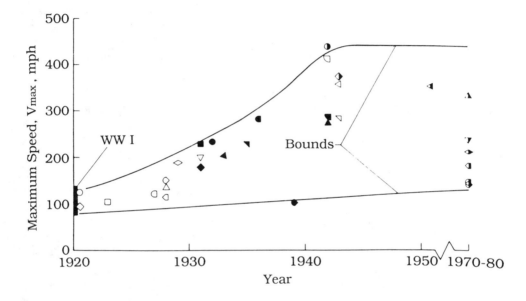

FIGURE 12-5.   Trends in maximum speed of propeller-driven aircraft. (From Loftin, 1985.)

had yet to be named. It would only be named "the Sound Barrier" after the development of a revolutionary aircraft using new means of propulsion. The rocket and jet aircraft of World War II led to the end of the first phase of aeronautical development, subsonic flight, and ushered in the next era of supersonic flight.

Table 12-3 lists the progression in maximum speed for propeller-driven designs from 1913 to 1969. As these aircraft flew faster, they encountered the effects of compressibility, which acted on lift and drag characteristics of the airfoil.[14] The increase in Mach number (speed of sound = 1.0 Mach number) from 0.4 to 0.8 sees an incremental lift coefficient value of only 0.2, together with a large increase (an order of magnitude) in drag. Aircraft approaching the sound barrier experienced instability and many times loss of control. Figure 12-5 shows the limits of propeller-driven speed.

Comparisons in thrust-to-weight ratios of early jets to high performance propeller-driven aircraft show an interesting disparity. A 3900 kg propeller-driven fighter with a maximum speed of 660 kph generates 2600 kg of thrust at 40 kph and only 410 kg at 660 kph or a thrust-to-weight of 6.75 (at 40 kph) and 0.12 at high speed. By comparison, an early jet of equal size with an 800 kg thrust has a thrust-to-weight of 0.25 throughout its range. These results imply a long takeoff roll for jets but the nearly constant thrust-to-weight ratio gives the jet tremendous advantage at the high-speed end of the flight range. That is, with equal drag area the jet would be expected to fly much faster. In actuality this proved to be true by as much as 250 kilometers per hour.

The Germans flew the first successful jet aircraft in 1939, the Heinkel (He) 178[15] (Figure 12-6). Later, out of desperate necessity, the German war effort produced the Me-262, first flown in 1942 and later to become the first operational jet fighter[16] (Figure 12-7A). Powered, by today's standards, by feeble axial flow Jumbo engines the Me-262 could reach 840 kph.[17,18] Two hundred of these aircraft were to see service. The origin of leading-edge slats for low-speed flight can be traced to the Me-262.[19]

All we have seen, fundamentally all early modern development in guided missiles and rockets derive from German inventions. Besides the "V-weapons" they developed the first practical rocket plane, the Me-166 or KOMET (Figure 12-7B). It flew in combat against allied B-17 formations in 1944. The small plane had advanced features like swept wings, a liquid-fuel Walther rocket engine mixed with primitive elements like the take off landing system of wheels and a skid (the wheels were jettisoned on take off). Maximum speed was 960 kph with a maximum burn of 2 min on the Me- 163C.[20] The Me-163C had a pressurized cockpit.

Boyne has called the Me-163 an example of mismatched technology—advanced engine and airframe with primitive armament, fuel and cockpit systems.[21] As engineering it was a classic case of tradeoffs—bargaining and sub-optimizing of things you want vs. what you can live without. Alexander Lippisch developed the airframe and Helmut Walter the engine. The fuel had two components—*C-stoff*, a mixture of alcohol, hydrazine, hydrate, and water, and *T-stoff*, hydrogen peroxide.[22] This mixture was highly explosive, corrosive and terribly dangerous to man and machine. Combined with the archaic landing skid, a loaded KOMET could not land without great peril of spontaneous explosion. Still, it flew and for all its deficiencies, it was operational by August, 1944, but in its brief career accounted for only 11 successful attacks.[23]

CHEMICAL ENGINEERING AND WAR

By 1840 Germany had advanced industrial chemistry to the modern stage of development. Germany traded on her position in World War I by shutting off her "fine chemicals" from the rest of the world. These variety of chemicals ranged from dyes to drugs and were costly and tricky to make. Further "intermediate" chemicals or specific organic chemicals were cut off as well. The U.S. began exploiting her own vast resources of coal, petroleum, phosphate, and sulphur to quickly offset the German advantage in explosives technology, acids and in "fine chemicals".

Coal and petroleum provided raw stuffs and power, phosphate provided fertilizer and nitric acid, and sulphur, from Louisiana, provided the all important acid, H_2SO_4. Coal-tar derivatives including aspirin (acetylsalicylic acid), barbital, novocaine, and phenacetin were duplicated by the U.S. in response to the WWI shut-off. "Hydrogenation" in the production of "vegetable fats" led to a butter substitute, while corn use led to alcohols and an important source of glucose other than sugar cane. Artificial fibers, cellulose that was actylated, produced "rayon". Cellulose and coal tar, subjected to heat and pressure, are the raw materials of plastic of Bakelite. Further research led to the long chain of polymers or giant molecules such as nylon.

Synthetic fuels were advanced by Germany's petro-chemical engineers. Forced to rely on her highly developed chemical industry, Germany in World War II, pioneered, from necessity, in the production of synthetic fuels and gas from coal. The U.S. has coal reserves of approximately 640×10 metric tons. This includes coal and lignite. This interesting parallel with World War II Germany and the U.S. continues in that wartime Germany utilized its "brown coal" for power generation and chemistry. This "poor" coal is very similar to

FIGURE 12-6.   The first turbojet aeroplane, the Heinkel He 178, 1939 (Smithsonian Institution).

our lignite. Lignite was first discovered to have fuel properties by the Greeks of antiquity who noted that the addition of fuel oil allowed it to support combustion and burn itself. Germany knew this and developed coal-fired plants that were underground and virtually undetectable due to pollution control systems such as electrostatic separators and chemical scrubbers.

U.S. scientists at Texas A & M University currently are translating the German archival documents on synthetic fuel production in World War II. This unique project envisages the exploitation of German processes in the critical area of energy production.

NUCLEAR POWER

It is in the 1930s that we see the most profound innovation of the 20th century—nuclear power. Based on Otto Hahn and G. Strassmann's discovery of nuclear fission in 1938, Niels Bohr and J.A. Wheeler (1939) completed a detailed model of fission where the nucleus splits with release of a neutron which can, in practice, start nuclear chain reaction, provided adequate amounts of rare isotope U^{235} is present. Following Albert Einstein, the energy available is this

$$E = \Sigma(M_o - M_i)c^2$$

where M_o and M_i are the masses of the original and product nuclei, at rest and unexcited, and c is the velocity of light. The first "pile" or nuclear reactor was built in the Stagg Field of the University of Chicago in 1942. Enrico Fermi was the designer of this device based upon his "Pile Theory". Briefly, a mass (or "pile") containing uranium is spread in some suitable arrangement throughout a block of graphite. Whenever fission takes place in this system, an average number of neutrons, v, is emitted at a continuous distribution of energy of the order of 1 MeV. [A neutron is the non-charged nuclei particle that charge balances the positive proton (hydrogen nucleus)].

There is a small probability, ϵ, that the neutron will be absorbed before its energy has been appreciably decreased. The leads often to the fission of U^{238}. This fission type has a very low probability. In the majority of neutron emissions, the energy of the "fast neutron" is dissipated due to collision with the carbon atoms in the graphite. It takes 6.3 collisions with carbon atoms to reduce the energy by a factor of e. It takes about 110 collisions to reduce the energy value from 1 MeV to the thermal value of 0.025 eV. This neutron can be absorbed by the resonance process in uranium

$$U^{238} + n \rightarrow U^{239}$$

or

$$U^{235} = n \rightarrow Fission$$

The X-10 Reactor at Oak Ridge (1943) was the first to produce an appreciable power level.

Its core was 24 feet cube of graphite blocks; 1248 fuel channels, each 4.4 cm square located in a 17 cm rectangular lattice. The fuel was 35 tons of uranium metal, 2.8 cm diameter, 10 cm long and jacketed with aluminum to prevent oxidation. The reactor was air cooled with 47,200 cubic meters of air/min flowing through the core channels (fuel was round in square holes of channels). At 3800 kW the fuel temperature was 245°C; graphite 130°C. The concrete shield was 2.1 meters thick.

CHAIN REACTIONS—BASIC PRINCIPLES FOR REACTOR DESIGNS

Maintenance of chain reaction depends on the case where at least one neutron created

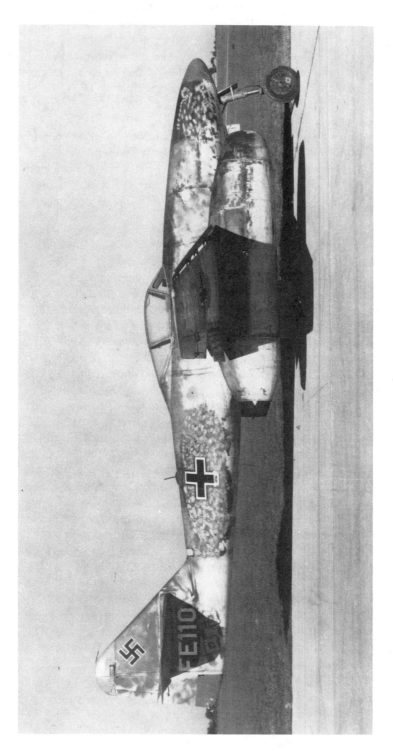

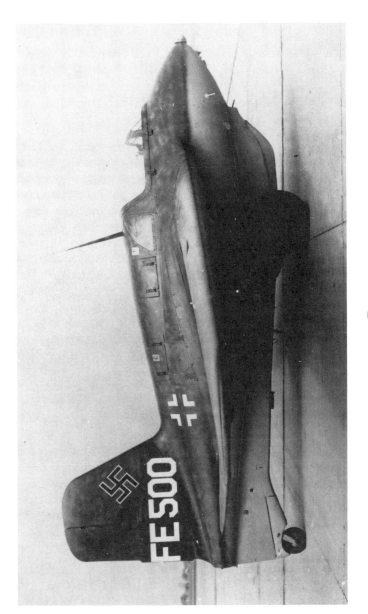

B

FIGURE 12-7. (A) The turbojet Messerschmitt Me 262 fighter. (B) The rocket-propelled Messerschmitt Me 163 fighter (Smithsonian Institution).

in a fission process caused another fission. This condition is expressed in terms of multiplication factor, k, where

k < 1, not self-sustaining chain reaction
k = 1, a self-sustaining chain reaction at steady state
k > 1, number of neutrons, fissions increase with each generation and
a divergent chain reaction results.

Fissile material is critical at k = 1; supercritical at k > 1.

Chain reaction growth is a function of τ; the average time between neutron generations is

$$\frac{dN}{dT} = \frac{N(k - 1)}{\tau}$$

which, after integration, is

$$N = N_o e^{(k - 1)t/\tau}$$

If τ is very short, and k >> 1, then there is an explosive chain reaction. At k = 1, τ = 10^{-3} s for thermal neutrons, and at k = 2, τ = 10^{-7} s. In a bomb, assembly must take place in approximately 10^{-4} s; detonation moves elements apart at 9×10^{-8} cm/s.

Earnest O. Lawrence developed electromagnetic separation methods of U^{235} from neutral uranium. It was, in effect, a giant mass spectrometer using a 462 cm (184 in) electromagnet. The Oak Ridge process was a diffusion process and produced large quantities over the macrogram (μg) amounts originally extracted by the spectrometer process. Plutonium (Pu^{239}) was also produced with the first weapons, being either U^{235} or Pu^{239}. The first successful fission weapon was the Trinity test bomb in New Mexico, July 16, 1945. The first fusion weapon, called MIKE, was detonated on November 2, 1952. The yield was 10.5 megatons. Edward Teller was called the ''Father of the H-Bomb''—H standing for the hydrogen used in the fusion process.

The formula for fission reaction is:

$$N = N_o \lambda \phi \sigma$$

where   N = number of fissions of U^{235}
N_o = original number of U^{235} atoms
λ = decay constant for U^{235} fission
Φ = neutron flux
σ = reaction cross-section for induced fission of U^{235}

Power is generated, today, in high pressure reactors which generate steam for the turbines thermodynamically and economically much like conventional power plants. More controversial is the so-called ''Fast'' or ''Breeder'' Reactor. As we saw in Table 12-4 of absorption processes, there is 77% probability of absorption of a neutron by U^{238}. This process yields highly fissile Pu^{239} by an (n,γ) reaction. Further Th^{232} can be converted to fissionable U^{233} by the same process. These reactors produce more fuel than they use in a ''breeder blanket''. They use liquid metal (Na, liquid at 90°C) in a closed loop to heat water for steam. It is projected that these reactors can increase the amount of fission fuel by a factor of 150. Nuclear power has created two parallel problems for today's engineers: (1) how to develop a viable alternative, and (2) how to dispose of radioactive waste being produced.

TABLE 12-4
Probabilities of Absorption Processes

Probability (%)	Process
3	Fast fission
10	Resonance absorption
10	Absorption by carbon
77	Absorption by uranium at thermal energies

The first problem is being addressed in both near and long-term research. Clearly petroleum and coal will be the primary source of energy for the next quarter century. Nuclear power may develop into a 25–40% component of the energy supply. Without designs like the breeder reactor being built (on-line) this projection may not be valid. Long-term solutions hinge on the development of thermonuclear or fusion power and gas technologies such as hydrogen—each, ironically, can use water as its fuel source.

FUSION

This process involves the combination of two light nuclei (1/1H) to form a nucleus which is more complex but whose rest mass is less than the sum of the rest masses of the original nuclei, i.e.,

Rest mass 4 hydrogen atoms = 4.03132 amu
Rest mass 1 helium atom = 4.00260 amu
Total 0.02872 amu (where 1 amu = 971 Mev)

$$0.02872 \times (971 \text{ MeV}) = 27.8 \text{ MeV}$$

The reaction is thus: (Proton-Proton)

$$^1_1H + ^2_1H \rightarrow ^0_1H + ^0_1e$$

$$^2_1H + ^1_1H \rightarrow ^3_2He + \gamma \qquad \text{Ionized Plasma}$$

$$^3_2He + ^3_2He \rightarrow ^4_2He + ^1_1H + ^1_1H$$

The temperature dependence of the reaction is evident with 10^{-7} cal/g s at 5×10^{5}°C to 3×10^{12} cal/g s at 1×10^{7}°C—at 3×10^{6}°C, 3×10^{7} cal/g s is produced.

DISPOSAL OF NUCLEAR WASTE

The second problem in nuclear power comes from the proper disposal of its waste byproducts—many long lived ($>10^5$ years). Ceramic and glass technology has been examined for containment and long-term storage of the more dangerous nuclides (Ni, Co, Sr). Two interesting studies—one by Corning Museum and another by the Department of Energy—have utilized a historical approach to disposal. The first study has examined the stability of archaeological glass and found Egyptian and Roman glass very stable over the last 4 millennia. The second study has examined how to mark a nuclear waste site for the same period (i.e., millennia). The results of the study isolated the following:

1. The disposal site should be marked with stone monoliths (ala Stonehenge)
2. Perimeter markers with comprehensive messages
3. Optional earthworks

TABLE 12-5
Some Impacts of Engineered Works

Engineering field	Impacts
Materials: Chemical and Synthetic Polymers; Carcinogens	
Ocean and marine	Well blow-outs, marine pollution, tanker disasters
Construction	Catastrophic failures due in part to flawed designs
Transportation	Air pollution, noise pollution
Aerospace	Air pollution, inefficient use of scarce fuels, pollution in near-Earth space
Chemical	Air, water polution; catastrophes like the Bhopal (India) disaster of 1984

4.   Central monolith, elevated with most detailed level of information
5.   The monoliths should be megaliths
6.   They should get attention, warn and describe and inform

SUMMARY

By the mid-20th century we have seen the accelerative impact of war on nuclear engineering. In rocketry the German effort, led by Dr. Wernher von Braun, exceeded all dreams and propelled the victorious allies (U.S. vs. Russia) into the "SPACE RACE". The "race" took on herculean proportions in 1957 with the "shot heard round the world"—SPUTNIK, the world's first orbiting satellite, launched by the Soviet Union. The effort of American engineers led to the first moon landing and the establishment of the computer as part of everyday life. World War I profoundly influenced the development of the aircraft. The use of chemicals in warfare and in vehicles was established by this war.

World War II led to advances in medical applications of chemistry (i.e., sulfa drugs, antibiotics). The development of the aircraft continued with the development of the first operational jet aircraft (Me 262). To offset Germany's advantages in undersea craft and air forces, Great Britain and the U.S. developed SONAR and RADAR. The profound development of nuclear power altered modern history forever.

All engineering fields of today must consider the environmental impacts of their works and products. Nuclear engineering is just the most obvious. Modern engineering technology must consider the real nature of "Damocles Sword" (Table 12-5). The impacts of the 20th century on engineered works are even more obvious than those of the 19th century.

NOTES

1. Chanute's 1896 glider design standardized biplane structure. Gibbs-Smith, H. *The Aeroplane, An Historical Survey of its Origins and Development*. Her Majesty's Stationery Office (1960), p. 36.
2. **Loftin, L. V., Jr.** *Quest for Performance*. NASA (1985), p. 44.
3. **Abernathy, W. J. and Utterback, J. M.** Patterns of industrial innovation. *Technol. Rev.*, **80**(7): (1978).
4. **Allen, F.** The letter that changed the way we fly. *American Heritage of Invention and Technology*, Vol. 4, No. 2 (1988), pp. 6–13.
5. **Kock, W. E.** *Radar, Sonar, and Holography*. Academic (1973), pp.51-66
6. **Kemp, P.** *The History of Ships*. Longmeadow (1988), p. 255.
7. Jean Baptiste Joseph Fourier (1768–1830), a French mathematician who taught at the École Normale and served as secretary for the Institute of Egypt, the scientific group that accompanied Napoleon during his expedition to Egypt. Fourier developed the theory on the conduction and flow of heat by use of Fourier Analysis—a theorem where any repetitive oscillation can be evaluated as a series (the Fourier Series) of sine waves of constant amplitude:

$$F(f) = \int_{\infty}^{\infty} g(t)(\cos 2\pi \, ft - i \sin \pi ft) \, dt \quad \text{(continuous)}$$

or

$$F(\vartheta) - \frac{1}{n} \sum_{r=0}^{n-1} g(t)(\cos 2\pi\vartheta r - i \sin \pi\vartheta r) \quad \text{(discrete)}$$

Navier extended Fourier's work beyond heat studies. Fourier served as prefect of the French section of the road to Turin, Italy and drained 80,000 square kilometers of swamps. (See figure, Bracewell, R. N. The Fourier Transform. *Sci. Am.* **260**(6): 86-95 (1989).

8. The primary comparison is between sonar's T_bA, the ratio of incident power to reflected power 1 yard from target, and radar's σ, the ratio of incident to reflected power at the target. So, when

$$\sigma \text{ (for radar)} = Ag(1\text{-}k)$$

and

$$T_bA = \frac{\sigma}{4\pi R_a^2}$$

than combining equations* yields

$$T_bA = \frac{\sigma}{4\pi} (R_a - 1)$$

*Various Quantities are A: Area of target; g: gain of reflected signal in directions of radar; k: absorption coefficient of target; T_b: sonar reflection coefficient. (After Collins, A.D.)

9. Britain's code name for their radar was "R.D.F."
10. **Goddard. R. H.** A method of reaching extreme altitudes. *Smithsonian Publication* No. 2540 (1919).
11. **Oberth, H.** *Die Rakete zu den Planetenräumen.* Munich: Oldenbourg Publishing (1923).
12. **Carvill, J.** *Famous Names in Engineering.* Butterworths (1981), pp. 61-62.
13. Ibid., p. 39.
14. **Loftin,** op. cit.
15. **Masters, D.** *German Jet Engines.* Jane's (1982), p. 9.
16. **Gibbs-Smith,** op. cit., p. 121.
17. **Boyne, W. J.** *The Smithsonian Book of Flight.* Orion (1987), p. 200.
18. **Gibbs-Smith,** op. cit.
19. **Boyne,** op. cit., p. 208.
20. **Masters,** op. cit., p. 115.
21. **Boyne, W. J.** *The Aircraft Treasures of Silver Hill.* Rawson Associates (1982), p. 213.
22. Ibid., p. 212.
23. **Masters,** op. cit., p. 113.

13 20TH CENTURY ENGINEERING: PART 2

PETROLEUM ENGINEERING—HEROIC DESIGN

The 19th century development of petroleum had proceeded at a comparatively languid pace compared to other technologies. One principal reason was the continued dominance of coal as a principal fuel for industrial processes.[1] Steam continued to power the Industrial Revolution as it spread beyond Britain. In 1850 coal supplied 10% of all energy consumed, but by 1885 this had risen to 50%.[2] Edwin Drake had made the first significant petroleum strike in 1859 but petroleum only accounted for 5% of the U.S. energy demands by 1890.[3] By 1945 it had risen, together with natural gas, to 45%.[4] Ninety percent of the early supply came from oil fields in a belt from Ohio to New York.[5]

This condition persisted until an extraordinary discovery was made in a coastal pasture 7 kilometers south of Beaumont, TX. That discovery was the Spindletop oil field, which erupted into life as a dramatic "gusher" spewing oil, gas, pipe and derrick parts across the Texas landscape.[6] The eastern monopoly ended overnight. This single well produced twice as much oil as all the Pennsylvania wells combined, with the field producing more oil than the rest of the world at that time (3.6 million barrels in 1901). The supply so exceeded demand that oil prices fell to two cents a barrel.

Innovation in drilling technology allowed the development of the Spindletop field. The rotary drilling rig was adapted from water-well drillers and in another innovation, "drilling mud" was pumped down the hole to strengthen the walls as the drill bit passed into the soft geologic formations of the Spindletop oil reservoirs. Beyond petroleum technology, the impact of cheap oil on rail and industrial engineering was even more immediate and pronounced.

The first oil-burning railway engine was delivered in 1901. By 1906 Southern Pacific engines were burning 2.6 million barrels a year at a cost savings of 40%.[7] Ships began to burn oil as tankers began to move the crude eastward. Oil occupied about half the space required by coal, leaving more room for income-producing cargo. The change to oil gave ships as much as 2000 tons more capacity.

Finds in California and Oklahoma made before 1920 eclipsed Spindletop, as its production fell. By 1940 additional fields in Texas and Louisiana had discovered 60% of the recoverable oil—88 billion barrels—in the U.S. Since World War II only one comparable field, Prudhoe Bay, with 10 billion barrels has been found.[8]

As fields declined, onshore exploration of the nearshore and then deeper waters began with the first offshore well drilled in 5.5 meters of water 19 kilometers offshore Louisiana in the Gulf of Mexico on September 9, 1947.[9] The effort mixed land-based drilling technology with marine systems for the first time, creating what is now termed "offshore technology". Thirty years later in 1978, Shell Oil placed the 46,000 ton COGNAC platform in water depths of 312 meters, drilling 62 separate wells.[10] With a tenfold increase in crude oil prices after 1973, uneconomic deposits in inaccessible places became economical overnight. One such place is the North Sea, the other the Arctic.

It is with the North Sea and the deep waters of the Gulf of Mexico that offshore engineering designs have become truly heroic in scale. Two principal systems are used today, with newer designs being developed as depths increase. The steel jacket platform (Figure 13-1) is the tied-beam or cantilever of Eiffel. The great jackets are constructed ashore in drydock, lying on their side, resting on flotation tanks. These jackets weigh 25,000 tons or more.[11] They are towed on launch barges weighing up to 10,000 tons and tipped into place,

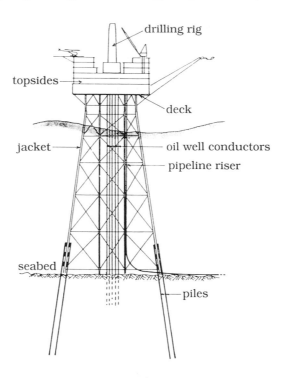

FIGURE 13-1.   Parts of a steel jacket platform.

while the flotation cells are flooded. Dead weights of 110,000 tons on some North Sea platforms resist the enormous sideways forces of the waves.

Using the lessons learned in the design of the COGNAC platform, engineers built and launched the world's tallest offshore structure, the BULLWINKLE platform[12] (Figures 13-2) in 1989. At almost 500 meters (1492 feet) this platform tops the tallest buildings. Unlike structures such as skyscrapers which are built on their erection site, the offshore platform must be fabricated on a shore facility, transported, and placed on its site. BULLWINKLE weighs in excess of 75,000 tons. Like all offshore designs it must resist a 20 meter wave and withstand fatigue stresses caused by millions of small ones. The platform has a period of 5 s and moves almost 2 meters at deck level. Under storm conditions the dynamic amplification is about 50%. The footprint of BULLWINKLE covers 1.8 hectares (124 × 148 meters). It tapers to 49 by 43 meters at the jacket top and expands upward to an 87 by 56 meter deck. Deck loading is up to 28,000 kips.

A major difference between COGNAC and BULLWINKLE, besides the size, lies in the techniques of construction and erection. COGNAC was fabricated in three pieces and erected in three stages over time. BULLWINKLE is a single-piece structure erected at one time. In fact, once towed to its site and the barge positioned, launching of the massive structure took only 80 s. Moving and erecting such a structure required computed-aided control systems at a scale not used in engineering before.

The second design used in offshore oil structures is the concrete gravity platform[13] (Figure 13-3). These platforms have developed in parallel with the steel designs. Their popularity, aside from the strength and durability qualities of prestressed concrete, stems in large part from their capacity to store large quantities of oil when oil cannot be pumped. One platform in the North Sea weighs over 230,000 tons. Gravity platforms have an added advantage in that their topside facilities can be completed before being towed to their offshore site. This advantage is less today with the larger capacity of offshore cranes.

New designs are being deployed in both the North Sea and Gulf of Mexico. The Tension

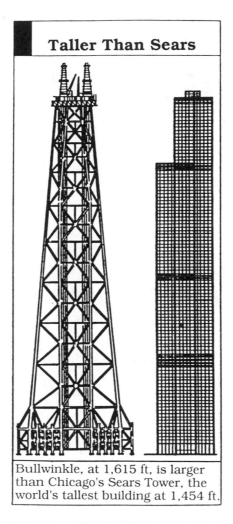

Taller Than Sears

Bullwinkle, at 1,615 ft, is larger than Chicago's Sears Tower, the world's tallest building at 1,454 ft.

FIGURE 13-2.   BULLWINKLE. (Civil Engineering)

Leg Platform (TLP), Conoco installed in the Hutton Field of the North Sea in 1984 and at a Gulf of Mexico Green Canyon site in over 300 meters of water (1988). The TLP has been developed for producing oil in very deep water. The platform is held in place by vertical legs kept in tension by the buoyancy of the platform resisting heave (vertical motion) and sway (horizontal motion). The Guyed Wire Platform design allows the structure to be held in place by stays or cables anchored to the sea floor. Again, this design is for very deep water. Such designs are truly heroic in scale. Every project pushes knowledge and expertise further. With the anticipated decline in onshore reserves of oil, the engineer will be called upon to meet and exceed the accomplishments of the present with even more innovative and economic designs.

The search for oil has led engineers to cope with the harshest of environments, the Arctic. In the mid-1970s exploratory drilling began in the Beaufort Sea.[14] The design of structures for oil recovery in the Arctic must consider the force of sea ice. The large lateral shear forces produced by ice must be compensated by designs unique to this requirement. Engineers have built artificial earth islands in 20 to 60 meters of water. The typical exploratory island is 100 meters in diameter with production islands 500–600 meters. They rise 4 to 25 meters above the sea ice. To bring oil from these fields pipelines such as the Trans-Alaska Pipeline from the Alaskan North Slope to Valdez on the hydraulic side have been built using technologies adapted or invented for these adverse conditions.

FIGURE 13-3.   Concrete gravity platform.

SUPERSHIPS

Oil and the economy of scale has led to the design of the largest ships afloat—the supertankers. They are classified in two classes: the VLCC (Very Large Crude Carrier) and the ULCC (Ultra Large Crude Carrier). The VLCCs range from 250,000 to 275,000 dead-weight tons with the ULCCs up to 500,000 deadweight tons. They are generally powered by steam turbines.[15] This reduces the risk of fire and provides steam to heat the oil for pumping. The steam drives pumps, capstans, and winches instead of electric motors. Most are single-screw ships using variable-pitch propellers to achieve economic performance. Their drafts are so great (18 meters or more) they can only use the deepest harbors. In shallow areas such as the Gulf of Mexico offshore facilities have been built to offload them far from land. In a way the supertanker is like the mythic FLYING DUTCHMAN by not coming to land.

FIGURE 13-4. Wreck of the AMOCO CADIZ. (National Oceanic and Atmospheric Administration)

These large ships do carry a curse. They are all potential environmental disasters by virtue of the spillage of oil. The wrecks of the TORREY CANYON (1967), AMOCO CADIZ (1978) (Figure 13-4), and EXXON VALDEZ (1989) created extensive pollution of nearby coasts. The TORREY CANYON spilled 117,000 tons of crude, while the AMOCO CADIZ and EXXON VALDEX spilled in excess of 200,000 tons each.[16, 17]

PETROLEUM ENGINEERING—RECOVERY OF A FINITE RESOURCE

On the approximately half-billion barrels of oil believed to be recoverable, over 30% were recovered by 1979.[18] Decline in U.S. fields is about 1.5 billion barrels, but recent estimates continue to point to a declining resource.[19] In response, engineers have turned to extensive modeling of the reservoir formations to develop techniques for enhancing the recovery of their oil and gas.

In the past the economic life of an oil reservoir typically ends when the solution-gas pressure no longer drives the spontaneous flow of oil.[20] On the average, only about 25% of the available oil has been recovered at this point. As offshore development was accelerated by the increase in oil prices to higher levels it has likewise made enhanced recovery techniques more economical.

Flow rates in reservoirs can be increased by either increasing the pressure gradient or by reducing viscosity. Enhanced recovery has a history of serendipity, almost as soon as commercial recovery of oil began in Pennsylvania. Holes in casing pipes allowed water from surrounding shallow aquifers to flow into the well. This "waterflood" increased pressure in the reservoir. Operators then began injecting water into their wells to offset the natural decline in reservoir pressure. It became an "underground" technique quite literally after Pennsylvania outlawed the practice, based on the erroneous assumption that the water would destroy the reservoir's productivity. By the 1930s waterflood was applied to Texas oil reservoirs and in Oklahoma's great Comanche Field's North Burbank Unit. Increase in efficiency due to waterflood techniques was a factor of 1.5 to 2.0 or recovery of 40% of the reservoir's oil.

The limit of simple waterflooding was reached due to grain and pore sizes in the reservoir. As most reservoirs are sandstone,[21] the viscosity differential between oil and water interacting with heterogeneous pore sizes can seriously reduce flow. To overcome this, the technique of *fracturing* has been evolved, where truck- or skid-mounted triplex pumps, powered by large diesels, force sand-laden fluids into the reservoir at pressures of thousands of kg/cm^2 to "fracture" the reservoir formation, thus allowing flow. To further enhance flow, surfactant solutions of water-soluble polymers, hydrolyzed acrylate polymers, and polysaccharides are injected as part of the fracturing process to reach the trapped oil and enable flow. Other techniques used today include thermal recovery, a technique again discovered accidentally in Venezuela, where steam injected into wells raises the reservoir oil temperature and lowers viscosity. A variant of this technique, used in large reservoirs such as in California, is called the "steam-drive" technique. Steam is injected into adjacent wells so the flow is to the primary well. The technique has raised recovery to as much as 70%. More expensive techniques such as the use of carbon dioxide under great pressure may become useful if oil prices make the practice economical.

A TUNNEL FOR THE AUTOMOBILE

To David Billington, civil works define civilization.[22] He sets public utilities at the center of civilized life, wherein water supply, transportation and collective shelter make today's world possible. War is certainly the enemy of civilization even though the catalogue of advances made during this century's conflicts appears to contradict this conclusion. Before, during and after the wars practioners of the useful arts of engineering have added to the quality of civilization as Billington defines it.

New York City became a major focus of activity in construction of transportation structures. Like London, Manhattan lies on a great river. In New York's case it is the Hudson and its branches. To connect Manhattan Island to the adjacent shores required the construction of more great bridges, like Roebling's Brooklyn Bridge, and tunnels that surpassed Brunel's Thames design. Where the Brunel tunnel originally carried only pedestrian traffic, New York's tunnels were designed for rail and vehicular use. In 1905 Alfred Nobel (1844–1914) drove rail tunnels under the Hudson River. The two tubes were over 7 meters in diameter and 1600 meters long. Under the shorter 1200 meter East River section four tubes were built. This was completed in 1909. The engineering problems in building tunnels under the Hudson were daunting. A tunnel had to be built through almost liquid mud and silt, necessitating careful control of the compressed air pressures in the construction. Too much pressure and the tunnel "blows out" and too little and it floods anyway. Nobel built on the previous work of the British engineers, Sir Benjamin Baker and James H. Greathead, who completed the two 4.8 meter diameter Hudson and Manhattan rail tubes in 1905.[23] Nobel and his co-workers dumped tons of clay on top of the tubes (over 300,000 cubic meters) to help prevent the deadly flooding during the 6 years of construction.

These tunnels provided for rail traffic but did nothing for the burgeoning vehicle traffic that supplied Manhattan with food and supplies, as well as workers.[24] The construction of two vehicular tunnels was begun in October 1920 following the design of Clifford M. Holland (1883–1924). His design was for twin steel-tube tunnels just over 9 meters in diameter and 2819 meters long, of which 1670 meters were sub-aqueous[25] (Figure 13-5a). The tunnel's tubes were about 4.5 meters apart, each with two one-way lanes and a walkway. The design feature that is most interesting is the ventilation system. It had to be designed to handle air flow over the tunnel's length and maintain safe levels of carbon monoxide.

Before the Holland Tunnel only two vehicular tunnels existed. These were the Blackwall and Rotherhithe (Thames) Tunnels under the Thames River at London. The Blackwall, opened in 1897, had an under-river length of 372 meters between ventilation shafts.[26] The

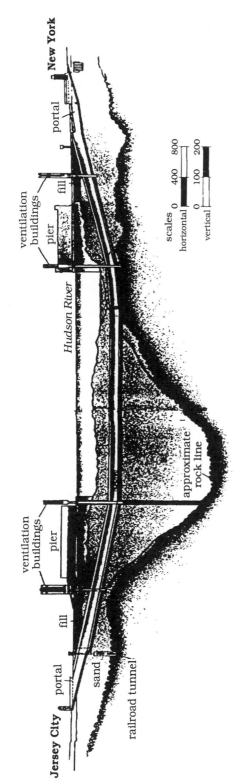

FIGURE 13-5. (A) Location plan of Holland Vehicular Tunnel; (B) Cross Section—one tube of Holland Tunnel. (Redrawn from *The Eighth Wonder*, 1927.)

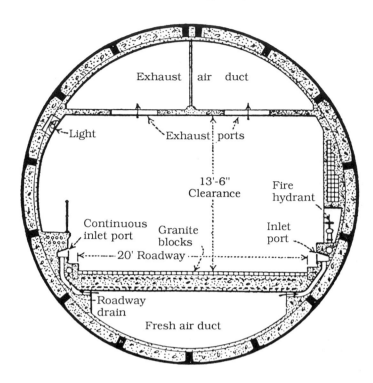

FIGURE 13-5b.

Rotherhithe was 478 meters. Both tunnels were ventilated by the natural movement of air through the shafts and end portals. Their vehicular traffic was roughly 100 automobiles per hour.[27] The Holland Tunnel was projected to carry 1900 motor vehicles per hour. This figure has risen to almost double the 1927 estimate.[28]

The length of the great Holland exacerbated the ventilation problem. At the Mont Cenis Tunnel through the French Alps the railroad engineers of the first train died of asphyxiation.[29] Carbon monoxide levels as a health problem were still the subject of experiment in 1920. The Bureau of Mines had conducted studies on exhaust gases to determine safe physiological levels. The exhaust levels for carbon monoxide ranged from 0.5 to 14%.[30] In tests on Yale University students it was concluded that four parts of carbon monoxide per 10,000 parts fresh air (4 in 10,000 parts of air) was a safe level. Today this limit is much more stringent: 1.25 in 10,000 parts of air.

To reduce CO levels to four parts per 10,000 would have required a longitudinal flow of air introduced at one portal from a nozzle at 421 kilometers per hour (280 mph). The velocity at points in the tunnel would have been 116 kph (72 mph). Clearly, this was prohibitive for vehicle use and the power needed to produce such a flow was excessive. Holland's solution was the use of air ducts adding and removing air at various points along the tunnel. The ducts ran above and below the roadway (Figure 13-5b). The fans used for the task were housed in four buildings, two on each end of the tunnel. This transverse flow did not produce high-velocity currents. Eighty-four fans, 48 blowers and 48 exhaust, were each powered by 200 hp engines.[31] Concentrations of CO on the roadway were measured at 0.5 to 9.4 in 10,000 parts of air.[32] Holland did not live to see his tunnel's completion. He died on October 24, 1924, 3 years before its opening by President Calvin Coolidge.

THE AUTOMOBILE

The first automobiles can be characterized as small carriages, powered by small engines, using pneumatic tires, built by bicycle mechanics. Engineering had already produced the diesel and Otto Cycle (1876) engines (see Chapter 12). Karl Benz, in 1886, produced the first commercially feasible automobile and sold his third vehicle at the 1889 Paris Exposition. His chief competition was Gottlieb Daimler (1885), an engine designer who sold the rights to Panhard and Lavassor of France. The 1891 vehicle of this firm placed the engine in the front of the chassis with the crankshaft parallel to the long axis of the car. This allowed the use of larger, longer engines.

Wilhelm Maybach, a colleague of Daimler, invented the carburetor. The first car to be produced in significant volume was the Oldsmobile (1901–1906), resembling more a horseless carriage than the French or German vehicles. However "low" in technology, the Oldsmobiles were affordable at $650 and were cheap to operate. Steering wheels replaced the tiller after 1901 and were generally placed on the left side. Other improvements rapidly followed in the water-cooled, force-feed lubricated engines, magneto ignitions, shock absorbers, bumpers, acetylene headlamps and vehicle tops.

Henry Ford built the successful Model N in 1906, followed by the even more popular Model T in 1908. The horsepower (hp) was 20 with extensive use of vanadium steels to lighten and toughen the frame. By 1927 15 million were sold for as low as $290. Autos sold in the U.S. at a higher volume due to higher per capita income, lower gas prices and better income distribution. Ford was more a practitioner of industrial engineering than automotive. The Model T was certainly not a better car than the Olds or other makes. Ford just made more of them at a cheaper price. To do this he adapted mass production technology to automobile construction.

On the race tracks of the world names such as France's Bugatti, Germany's Auto Union and Mercedes Benz makes became legends in the 1930s. In 1934 the Nazi government of Germany had Dr. Ferdinand Porsche develop a small "people's car" (Volkswagen). By 1938, the VW38 was introduced. In 1949 the Bug was built for non-military purposes. The 725 kilogram car had a flat air-cooled engine in the rear, producing 36 hp. Over 20 million have been built, making it a mid-century version of Ford's earlier success.

Chrysler featured four-wheel hydraulic brakes in 1925 (already introduced in the 1920 Duesenberg, also with a straight-eight engine). Chrysler made the first American aerodynamically designed mass-produced automobile in 1934. This was the Airflow. The car was a flop economically due principally to its unusual looks. The car was the brainchild of Carl Breer.[33] It had an aerodynamic drag coefficient (Cd) of 0.5–0.53. The 1936 VW had a Cd of 0.49. To appreciate this value, one obtains a Cd for a sphere of 0.10 and 1.17 for a flat iron plate. Airfoils such as wings yield Cds of 0.05 but they are designed to lift as well as slice through the air. The automobile must move efficiently through the air yet stay firmly on the ground. The Airflow did both due in large part to its heavy body and frame. Still, it could cruise at 130 kph in relative comfort. Octane ratings and tetraethyl lead were in place by 1926 and higher compression engines (4.5:1). The modern V-8 engine was introduced by Cadillac in 1914. The first practical low-priced V-8 was the 1932 Ford. Chevrolet's 1919–1923 experiment with an air-cooled engine was the last attempt by an American manufacturer to pioneer a major change in engine design. The engine was only produced in small numbers (759). Only 100 were ever sold. After this failure, General Motors (GM) developed a market strategy that discouraged innovation. Control passed from engineers to accountants. Constant upgrading of product really means planned obsolescence with cosmetic changes. Body styles changed with little engineering improvements. An example is the interchangeability of the parts of a 1929 Chevy six cylinder with a 1953 model. GM called the Cadillac V-8 (1949), with high compression (7.5:1) overhead valves, a major development but it was hardly revolutionary. What the 1949 engine did start was the era of the American "Dinosaur"—big in horsepower, low in fuel economy. The 1949 Cadillac did achieve 20 mpg, compared to the 1973 average of 13.5 mpg.

European imports met the demand for fuel economy while quality in U.S. cars fell. By the 1960s American-made cars had an average of 24 defects per car, many safety related.

The classic example of this is the 1960–1964 Corvair, where the corporate cost-cutting omitted a $15 stabilizer bar despite evidence that the car would oversteer dangerously. Mass production gave advantages to large companies due to economies of scale. In 1908 there were 253 auto makers in the U.S.; in 1929 there were 44; and in 1988 there are three. By 1970 European and Japanese manufacturers were producing technologically advanced, well-built, fuel-efficient lower-priced cars.

Innovations such as radial tires, independent suspensions, disc brakes, fuel injection, and front-wheel drive were all introduced abroad. Mavericks such as Preston Tucker and Raymond Lowey persisted but their designs have failed the economic test. American contributions such as the tubeless tire, alternator and the torsion-bar suspension were minor by comparison.[34] Similarly, the Europeans and Japanese produced turbochargers, stratified, multi-valve engine systems, and rotary engines. The U.S. world market share dropped from 76.2% in 1950 to 19.3% in 1982. In 1980 Japan replaced the U.S. as the largest manufacturer in the world.

U.S. manufacturers have begun to meet the requirements of a changed world market. New designs reflect increased use of new lightweight, high-strength materials and aerodynamics. In the efforts to achieve increased fuel economy, engineers have found that aerodynamic designs that reduce the Cd by 10% result in a 3.5% increase in economy. Likewise, a 10% reduction in weight results in a 2.5% increase in fuel economy. The most expensive way to achieve increased economy is improved engine design. The average weight per car has dropped from 1750 kg in 1975 to 1230 kg in 1985. Computer-controlled systems are becoming the rule in engines and suspensions. The potential of alternate fuels such as hydrogen is promising.

On the positive side, automobiles are safer, more fuel efficient, better built and less polluting. This renaissance in technological innovation may indeed replace the planned obsolescence dogma that almost destroyed the U.S. auto industry. The future of the automobile is rosy. Unlike public transportation systems, it is convenient. In many urban areas, it is safer. Modern women spend 8 h a week in the car as their own delivery service.[35] Goods and services have expanded out of central urban centers in response to this greater consumer mobility. Modern transportation infrastructures reflect this decentralization that is both the product of and succor for the automobile. Mass transit is making inroads in urban corridors, principally rail for commuters. "Super trains" compete with aircraft for medium-distance transit. Still, the car looks to have molded itself into modern society for years to come. The one safe prediction we can make, given the history of the U.S. auto industry, is that Detroit will never build a Ferrari.

RAILROADS *REDUX*

The evolution of railway engineering has been a story of rapid worldwide expansion after the early successes in Great Britain by the Stephensons and Brunels (see Chapter 12). This development increased from 1860 to the first World War. After this time, expansion slowed and has been in retreat ever since due to the competition of modes of transportation. In Europe trains have maintained a fairly consistent infrastructure for both passenger and freight traffic, while in the U.S. only a semblance of passenger service remains. This situation has resulted from the expansion of highways for motor vehicles which carry people and goods. With the success of the Douglas DC-2/3 in commercial aviation the economics of carrying passengers over long distances shifted from the railroads for good. This was not even considered in the halcyon days of early 20th century railroading. The steam locomotive had evolved in a steady progression to a relatively simple, powerful machine for either tractive effort (freight) or speed (passenger—service). In 1938 the British engine, MALLARD, set the world record for steam rail locomotive at 206 Kph (126 mph).[36] This represented the culmination in the improvement in efficiency of the steam locomotive design which began with the ROCKET in 1829. Thermal efficiency as measured in thermodynamic terms is simply the difference in temperature of the working substance (steam, air, etc.) before and after its use. As Carnot discovered, it is therefore desirable to start with the highest initial temperature and end with the lowest final temperature. Because of the shape and size of rail locomotives efficiencies never reached much above 8%.[37] By 1935 American railroads steam's successor was in service. The ZEPHER was a hundred-mile-per-hour streamliner powered by a revolutionary marriage of prime movers—the diesel and the electric motor.

Rudolph Diesel's engines were poorly suited to rail use. The early models had poor efficiencies and power-to-weight ratios of one horsepower to 300 pounds. These engines used most of their power to move themselves, much less anything else. It was for Charles Kettering to design, and General Motors to build, the first successful rail diesel engine.[38] Kettering invented the automobile self-starter, ethyl gasoline, and variable-speed transmission as well as the two-cycle diesel. This diesel design combined the four cycle's intake and exhaust strokes into the power stroke, making a smoother, faster engine. He combined his lighter design with the Salisbury unit fuel injector which removed the heavy fuel system plumbing by separating each injector for each cylinder. Kettering's engines were built of a then-new chromium, manganese, silicon-alloy steel called Cromonsil, which further reduced the weight. His eight-cylinder, two-cycle diesel produced 75 hp per cylinder at a ratio of 1 hp per 22 pounds, an order of magnitude difference over the older four-cycle engines.[39]

The diesel did not drive the wheels as the steam engines did. Instead, the rotational energy was taken from the engine's shaft and converted into electrical power by a direct-current (DC) generator. This current fed series-wound, DC traction motors which could be reversed on downgrades, providing dynamic braking and reducing wear on brakes. Combined with Westinghouse's automatic air brake (1872), the diesel locomotive replaced the heavier, expensive, less-efficient steam engines. By the 1950s steam engines were no longer manufactured in the U.S. and by 1961 28,500 diesels were at work, hauling twice the tonnage of 50,000 steam locomotives at half the cost.

Twentieth century advances in track standardization, brakes, control, and coupler systems, as well as automatic traffic control did not stave off the demise of real passenger service in the U.S. Railroads concentrated their service into freight hauling and even then faced competition from water-borne and highway systems. Still, Europe had the transportation infrastructure which it never lost in place after World War II. Post-war developments continued in passenger train design but it was for the Japanese to bring passenger rail service back as a viable alternative to air or highway travel for short-to-medium distances (100–500 kilometers).

The development of high-speed rail by Japan National Railways in the 1960s spurred new interest in passenger trains. Coupled with soaring fuel costs, the economics of rail travel became more appealing for moving people. The new trains, called "Bullet Trains" or "Shinkensen", were highly aerodynamic, high technology designs that combined roadway and train into a system. The powerful electric motors are driven by electrical overhead grids, removing the heavy diesels from the vehicle, while the rails and roadbed are specially designed and monitored by computer. In 1981 the French "TGV" (train gran vitesse) reached a maximum speed of 380 kph (236 mph) while averaging 250 kph (155 mph). By 1989 this top speed was raised by a new TGV to over 400 kph. Newer designs by the Japanese and German rail engineers dispense with the wheels altogether, using magnetic repulsion to "levitate" the train above the tracks. Prototypes have already exceeded 500 kph over short distances by removing the frictional resistance that had plagued railroad engineering since the ROCKET.[40]

The idea that a train could be raised above a track and propelled by magnetic means was German in origin, harking back to the 1930s where German engineers wrote papers on the theory of "attractive levitation". In a world not unlike Fritz Lang's 1926 "Metropolis" these trains would rush along guideways using electro-magnetic force. World War II interrupted this development but the idea remained. However, it was not for Germany to build the first successful maglev; this honor fell to the U.S.

Henry H. Kolm, an electrical engineer at MIT, built the MAGNEPLANE, a scale model maglev in 1973. Tested at Raytheon, the model system demonstrated a functional maglev over 240 meter track using super conducting magnets, block-switching technology and remotely sensed vehicle position along the guideway. Even though the MAGNEPLANE was an engineering success it was not funded and the leadership again passed to the Germans, with the Japanese engineers of the Japanese National Railways in hot pursuit. Germany had several maglev prototypes by the late 1970s and tested a full-scale, passenger-carrying maglev, the TRANSRAPID 05, in 1979. The JNR engineers produced a 500 kph maglev (without passengers) in 1978. By 1981 the Japanese had a maglev that combined passenger capacity and speed. The Germans were building TRANSRAPID 06, a 500 kph maglev capable of carrying 200 people.

SUBWAYS

The British developed the world's first subway, begun in 1860. Its tunnels were in reality cut-and-cover construction, with much of the line above ground as the motive power was provided by locomotives used on the British surface railways. They did, however, burn coke and condensed their own steam by 1863.[41] Deeper, later lines were tunneled and used electric power (1905). Paris followed the British lead, beginning construction of "Le Metro" in 1898. Its first sections were built just under the existing streets by a modified version of the cut-and-cover method. The chief engineer, Fulgence Biènvenue, used a pilot gallery begun top down and expanded outward from that axis to create a floored and walled tunnel. Where the London subway sought to relieve traffic congestion, the Metro was built to solve a problem in sanitation. The garbage men could not get through. The Metro was electrified from its inception.

These two examples become the models for 20th century subway systems. New York's subway was begun in 1900 with 34 kilometers complete by 1904, eight of which were electrified. William Barclay Parsons (1859–1932) was chief engineer. Parsons used steel bents for columns and beams with concrete between. During the construction of the subway the workers came across the remains of the first New York subway built by Alfred Ely Beach in 1870. It was more a full-scale model rather than a public transportation system. Only 95 meters in length, Beach's innovative design ran on track in a circular tunnel using pneumatics for the motive power. Utilizing a large fan to exhaust air ahead of the single car, Beach's subway car became a piston in a cylinder, successfully carrying passengers on numerous demonstrations until the financial crash of 1873. Afterward, Beach went on to publish *Scientific American,* sealing his subway to be discovered by Parsons' later works.

The year 1931 saw the start of one of the most important subways of the 20th century—the Moscow Subway. The construction of this subway posed problems in tunnel construction not encountered by builders of the London, Paris or New York systems. The soil under the Soviet capitol was so unstable that it was nearly impossible to make a drive even with shields. The chief engineer, P.P. Zarembo, and his assistant, Nikita Krushchev, utilized a chemical process for hardening the soil. A grout of silicate of soda ("water glass") was injected into the ground up to 10 meters, using perforated pipes. This hardened, facilitating tunneling which could then be lined with concrete. The advance was rapid (1 meter/shift) but costly. Workers who worked 48–72 h in the caissons often developed the bends as the Soviets used up to 2.3 atm pressure. Nonetheless, the first line opened May 15, 1935. The Moscow subway stations have their air changed four times an hour, stations are cleaned each morning, and the tunnels are washed with high pressure hoses once a month. High speed escalators (1 m/s) were first used in this system.

For the future an example of subway design has been suggested by Lawrence K. Edwards in 1967. This engineer and designer of the POLARIS missile proposed a "gravity vacuum" train for the Bay Area Rapid Transit (BART) system. While not built, the idea plays on Beach's earlier 19th century New York model, using a vacuum between stations and gravity to accelerate trains and allowing grade and air pressure to slow them at stops. One can visualize an undulating route like a sinusoidal wave train with stations at the crests and gravity "wells" at the troughs. Large fans at the ends and along the route would provide the vacuum in the tubes. Robert M. Salter (1978) of Rand Corporation proposed a coast-to-coast "subway" using gravity and magnetic levitation. His design would provide travel between New York and California in 54 minutes.

FUSION POWER

To ignite a hydrogen bomb, a "trigger" of a fission bomb is required. The unrestrained reaction, such as outlined above, produced tremendous energy, several orders above the magnitude of a fusion device. While moderation of fission is relatively straightforward, at sub-critical levels using graphite or water to absorb neutrons, the control of a fusion reaction is another matter. After a quarter century and several billion dollars no practical fusion reactor exists. Due to the temperatures necessary (approx. 10^9 degrees C), containments by ordinary materials is unworkable; hence the need for more innovative means.

MAGNETIC CONFINEMENT

The original British torus design, ZETA, of the 1950s did not achieve fusion. The U.S. design, the "C-Stellarator" (ca. 1958) was similar, being a toroidal design only twisted like a pretzel. It did not achieve the required temperature (100×10^6 degrees) at the right density (100×10^{12} nuclei) for the required time (approx. 1 s) either. The toroidal designs of the early fusion research programs could not prevent instability or "kinks" in the plasma which could touch the wall of the containment vessel, causing an explosion.

A leading Soviet fusion researcher, Lev Artsimovitch, in the 1960s, spurred development of the "Tokamak" ("Tok"-toroid, "mak"-magnetic, and "kam"-chamber). The Russians used large magnetic fields to control the kinks in the plasma. The adoption of this design by the U.S. and others began in 1969 after Artsimovitch delivered a series of lectures at MIT in the spring of 1969.

Containment of fusion plasma of energized nuclei and electrons has to maintain the plasma in both density and temperature. The natural path for a plasma is expansion and subsequent cooling (the Second Law of Thermodynamics). Confinement by magnetic fields has led to the design of a fusion reactor that resembles a doughnut. In this reactor the plasma is confined as a crescent-shaped ellipsoid. These reactors are "Tokamaks". The objective of this design is to maximize plasma pressure (hence temperature) at moderate magnetic levels achieving "breakeven"—the production of power such that the plasma is hot enough for self maintenance.

Current "Advanced Tokamak"[42] designs are poised to reach ignition temperatures at 4.7 keV and fusion reactions (D-T). Twenty keV temperatures have been reached.[43] The deuterium-tritium plasma to be produced will produce a flux of 14 meV fusion neutrons, intense enough to keep the plasma running and yielding a heat output capable of secondary power generation. Breakeven here is defined by the term, Q, on ratio of fusion-energy output to plasma-heating input.

INDIRECT OR LASER FUSION

Direct drive laser inertial fusion is inherently a simpler way to produce fusion energy. A pellet of fuel is illuminated by a high-energy laser or ion beams. The fuel is heated with the hot blowoff, driving the rest of the fuel inward to high density by a spherical rocket effect. The fuel chamber requires no magnets and no high vacuum. Simple in principle, inertial fusion requires: (1) excellent symmetry of illumination, (2) control of hydrodynamic instabilities, (3) minimal preheat of cold fuel by superthermal electrons, and (4) high rocket implosion efficiency.

Two approaches, direct and indirect drive, differ in whether the laser or ion energy is directly applied to the pellet or reflected as soft X-rays onto the fuel. Direct drive has problems due to fundamental problems with large, high power lasers in terms of optical imperfections in beams. Superthermal electrons plague both approaches. These originate in laser-plasma instabilities. Shorter laser wave lengths have essentially solved this.

X-ray conversion efficiency is lower than expected in the indirect technique (50–70% vs. 90%). Beam smoothing may improve this conversion efficiency. The optimum laser may be the KrF in the kilojoule power range since it has a short wave length (to reduce superthermal electrons) and a large bandwidth (for smoothing control of the beam).[44]

Both experimental Tokamak and laser-fusion systems have produced plasmas that approach the magic "breakeven" point in the terminology of fusion economics. This is the level where power produced equals power required to produce it. Neither approach has reached it as yet, much less produced an economically useable surplus of power.

ENGINEERING SPACE FLIGHT

Programs in the development of rocketry and instrumented exploration of space had proceeded in the U.S. and the Soviet Union since the end of World War II. Both nations had benefitted from the captured German plans and launch vehicles, as well as the incorporation of the actual persons into their space programs. Both nations were developing scientific satellites but no one could have gauged the impact of the launch of the Soviet satellite, SPUTNIK I, on October 4, 1957. It has been called a "technological Pearl Harbor" by American commentators.[45] The launch of SPUTNIK did demonstrate the Soviet Union's superiority in long-range, "heavy-lift" rockets. It did not presage any overall eclipse of Western technology as the alarmed media suggested.[46]

The Soviet launches were designed under the direction of Sergei Pavolovich Korolev (1900–1966) and his top assistant L. A. Voskrensky.[47] The "secret" of the Soviet success lay principally in the multi-engine booster, which was little more than a package of smaller, less-powerful rockets combined into a larger, more capable design. Following their lead and that of American space engineers such as Von Braun, Willie Ley and Hans Haber, the ante was quickly up from satellites to manned flight.[48] Even though the U.S. had concentrated on intercontinental ballistic missile design (ICBM), the disparity in lift capacity seen in the satellites, SPUTNIK at 83.5 kilograms, was still evident.[49] In fact, the 1964 VOSKHOD, a second-generation Soviet spacecraft replacing the one-man VOSTOCK, weighed in at 5300 kilograms, almost the weight of the U.S. APOLLO spacecraft of 5 years later. The U.S. National Aeronautical and Space Agency (NASA) formed out of the National Advisory Committee on Aeronautics (NACA). In existence since 1915, NACA was the service organization which conducted extensive scientific laboratory research in aeronautics.[50] Its mission was aeronautical research and the development of true aerospace vehicles. Implicit in this was the charge to "catch the Soviets". Before the MERCURY series of sub-orbital manned flights could begin in earnest, the U.S. was one-upped by the Soviets again. The VOSTOK series put men into space with the launch of Yuri Gagarin on April 12, 1961. Challenged, the U.S. response was to develop a program to place a man on the moon, and did so successfully July 20, 1969. Between these watershed events was an era of almost unparalleled advance in technology for space flight.

To fly in space engineers had to contend with deficiencies in computer capability, poor understanding of exact orbital motion (particularly that of the lunar orbit), unsophisticated guidance and tracking systems (ballistic missiles required little compared to that needed for orbital missions), materials for thermal protection of spacecraft, and landing systems.[51] Jack S. Kilby (1923–) and Robert N. Noyce (1927–) working independently solved the computer problem by developing the silicon microchip. The science of astrodynamics was born out

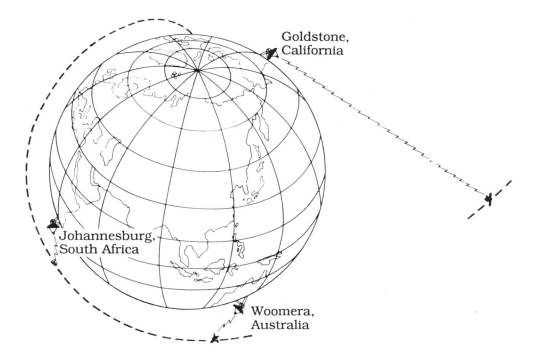

FIGURE 13-6.   Location of Deep Space Network stations. (Redrawn from Nicks, 1985.)

of necessity to compute space trajectories.[52] New materials such as teflon and thermal-resistant ceramics were discovered. In tracking, the development of global communications employing microwave and satellite technology led to networks like the Deep Space Network (DSN) with station antennas 120° apart (Figure 13-6) which recovered 99% of the 17,000 images transmitted by VOYAGER I at Saturn and determined its position within 337 kilometers of that planet.[53] New technology in liquid and solid fuel booster rockets led to the great lift capacities of the lunar rockets.

The hydrogen-oxygen liquid propellant technology is based on the equation:

$$O_2 + 4H_2 - 1.98\ H_2O + 1.97\ H_2 + 0.06\ H + 0.03\ OH + Heat$$

was combined with advances in variable thrust motors that could be aimed by swivels made for smoother flight, navigation and landing. The launch of a rocket has been termed a "controlled explosion". Ballistic missiles since the German V-2s were really atmospheric devices. The future designs required that they would never again see the atmosphere after launch and would survive for extended periods in deep space.

Advances in solid state propellants required a great deal of engineering devoted to the shape of grain, the configuration of the combustion chamber, and the control of burning of the propellant.[54] Modern solid propellants belong to one of two classes, double base or composite grain propellants. Typical compounds that make up the first are nitroglycerin, $C_3H_5(ONO_2)_3$, and nitrocellulose, $C_6H_7O_2(ONO_2)_3$. Both fuel and oxidizer are contained in the molecules. Nitrocellulose adds strength to the grain and nitroclycerin the high-performance burning component. Their mixture resembles a putty. The composite grain propellant is a mixture of rubbery polymer mixture of a fuel (polybutadrene) and a crumbly oxidizer such as potassium nitrate (KNO_3) or ammonium perchlorate (NH_4CrO_4).

The combustion of the solid fuel propellant produces an insulator effect at the fuel face and a resultant gas stream. The flow of the gas stream is described by the rate equation:

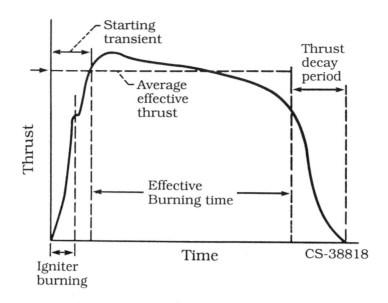

FIGURE 13-7.   Typical thrust-time diagram. (From Luidens, 1971.)

$$W = A_b r_\rho$$

where A_b is the total area of the combustion, r the rate of combustion and ρ the density of the gas. The thrust is given by $F = WIsp$, where Isp is the engine's specific impulse. The performance of the solid fuel engine can be evaluated by use of thrust-time diagrams such as Figure 13-7. Unmanned flights were sent to the moon, Mars, and Venus. Manned vehicles flew first to orbit, then beyond.

Successful examples of U.S. unmanned spacecraft included PIONEER 1 which discovered the Van Allen radiation belts encircling the Earth; the RANGER series (Nos. 7, 8, and 9) which first photographed the lunar surface (1964–65); the SURVEYOR series (Nos. 1, 3, 5, 6, 7) which soft-landed on the moon transmitting over 17,000 photographs and simple analyses of the surface soil (1966–68); the MARINER series [No. 2 Venus flyby (1962), No. 4 Mars flyby (1965), No. 5, Venus flyby (1967), No. 9, Mars Orbit (1971), No. 10 Mercury and Venus flybys (1973)]; the VIKING series, Nos. 1 and 2 which landed on Mars (1975); and the VOYAGERS 1 and 2) which flew by the Jovian, Saturn, Uranium, and Neptunian systems (1977–1989); and PIONEER 10, first to leave the Solar System (1983) after a Saturn flyby (1973).[55, 56] At some time after the death of Korolev and the loss of a Soviet cosmonaut (Vladimir Komarov, April, 1967) the two space powers took different paths.

SOYUZ I, the doomed first mission of that series, may have prevented the Soviets from flying a circum-lunar mission before APOLLO 8 (December, 1968)[57] SOYUZ as a program did evolve into a concerted orbital space station-endurance study which continues today. APOLLO went on to put 12 astronauts on the lunar surface before its end in December 1972. The SOYUZ program flew 15 orbital missions and launched the SALYUT I space station. The U.S. SKYLAB (1973–74) logged 513 days during three separate missions.

The 56,700 kilogram SKYLAB brought the two space powers to parity in lift capacity. The Soviets have gone on to build ENERGIA, now the successor to the SATURN in terms of thrust. The American space program concentrated on the construction of a reusable space plane or ''shuttle''. About the size of a small commercial jet liner, the SHUTTLE consists of three elements—the orbiter; an external liquid fuel tank; and two solid fuel booster rockets (Figure 13-8).[58] It is part aircraft, part missile. Flown successfully in 1981, it was the

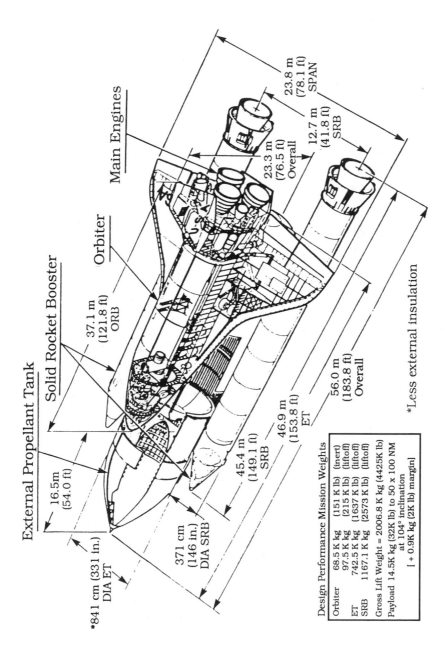

External Propellant Tank

Solid Rocket Booster

Orbiter

Main Engines

*841 cm (331 in.) DIA ET

16.5m (54.0 ft)

371 cm (146 in.) DIA SRB

37.1 m (121.8 ft) ORB

46.9 m (153.8 ft) ET

45.4 m (149.1 ft) SRB

56.0 m (183.8 ft) Overall

23.8 m (78.1 ft) SPAN

23.3 m (76.5 ft) Overall

12.7 m (41.8 ft) SRB

*Less external insulation

Design Performance Mission Weights

Orbiter	68.5 K kg	(151 K lb)	(inert)
	97.5 K kg	(215 K lb)	(liftoff)
ET	742.5 K kg	(1637 K lb)	(liftoff)
SRB	1167.1 K kg	(2573 K lb)	(liftoff)

Gross Lift Weight = 2006.8 K kg (4425K lb)

Payload 14.5K kg (32K lb) to 50 x 100 NM at 104° inclination
[+ 0.9K kg (2K lb) margin]

FIGURE 13-8. The Space Shuttle System (NASA).

complex vehicle design that produced the worst manned-flight tragedy to date with the loss of CHALLENGER in 1985. This catastrophic destruction of the vehicle was due to the failure of an O-ring seal on the solid fuel booster which ignited the external hydrogen fuel tank. Since that time it has reflown and has competition from shuttle designs from Russia, France and Japan. The shuttle vehicles only fly to low-earth orbit. The U.S. mission is to carry components of a permanent, manned space facility. From this platform missions to the moon and Mars are contemplated. The "space race" has produced supercomputers, materials such as new polymers, metals and ceramics, as well as a fine technology of micro-size and digital processing. It is difficult to gauge the impact of space flight on engineering as a whole at this juncture. We are too close to the era in historical terms; indeed we are still in it. Let us just say that projects like BULLWINKLE probably weren't possible without the advances made in engineering due to the challenge of space.

BRIDGES—*ART REDUX*

"Firmness, utility, pleasant appearance."—Vitruvius

In the 20th century bridge design has brought engineering to a level that has been called "structural art".[59] Othman Ammann (1879–1965) used suspension (George Washington, 1931; Verazzano, 1965) and arch (Bayonne, 1931) in designs surpassed the Roeblings, Streuss (Golden Gate, 1937) and Eiffel. Concrete construction in late 19th century advanced with the first production of artificial cement by Joseph Aspdin (1779–1855), called "Portland Cement". Reinforced concrete construction began in the late 1800s with Francois Coignet and T. Hyatt. Hyatt, particularly, used the reinforcing in the tensile zone of the beams. Reinforcing gave concrete (always high in resistance to compression) great resistance to tensile forces. Accidentally discovered by a Paris gardener, Joseph Monier (1823–1906), who received a patent in 1867, by 1875 he built a small reinforced concrete bridge using the very thin arch.[60,61] Lost to engineering history were the earlier works of T. Hyatt (1816–1901), who was originally a lawyer, and who developed reinforced concrete beams. These beams were correct in placement of the reinforcement steel in the tensile zone bent upwards at the supports, and anchored in the compression zone by vertical steel stirrups.[62] He also discovered the equality of thermal expansion coefficients of iron and concrete, which contributes to the high fire resistance of the medium.[63]

The first engineer to examine the theoretical nature of reinforced concrete structures was the German, M. Koenen (1849–1924), who stated the relationship of tensile and compressive forces in statistical terms.[64] The first successful bridge in this material was the Langweis Viaduct, with a 100 meter span, built in 1912 by E. Mörsch, which combined strength (it was a railway bridge) and an elegant slenderness. The fullness of range and possibilites of concrete bridge design was to await the brilliant Swiss engineer Robert Maillart (1872–1940) who followed Carl Culmann (1821–1881) and Wilhelm Ritter (1847–1906) and built on the tradition of daring design such as the Langweis Viaduct. Maillart used the hinged and deck-stiffened arch designs in dramatic ways, spanning up to 90 meters in a springing (e.g., Salginatobel, 1930).

Christian Menn (1927–) has followed Maillart, using concrete as a media, but pre-stressing his structures to create even more dramatic spans such as Ganter Bridge (1982, a pre-stressed cable-stayed design) (Canton Valais) whose main span is 174 meters with a column height of 150 meters. Significant bridges of the 20th century by these and other designers are listed in Tables 13-1 through 13-7.

TABLE 13-1
The Leading Suspension Bridges

Bridge	Main span (meters)	Completion date	Location
Hull-over Humber	1410	1982	Great Britain
Verrazano Narrows	1298	1964	New York, NY
Golden Gate	1280	1937	San Francisco, CA
Mackinac Straits	1158	1957	Mackinaw City, MI
George Washington	1067	1931	New York, NY
Salazar	1013	1966	Lisbon, Portugal
Forth Road	1006	1964	Queensferry, Scotland
Severn	988	1966	Beachly, England
Tacoma Narrows II	853	1950	Puget Sound, WA
Kammon Straits	712	1973	Kyushu-Honshu, Japan
Brooklyn	486	1883	New York, NY

TABLE 13-2
The Leading Cantilever Bridges

Bridge	Main span (meters)	Completion date	Location
Quebec (Railway)	549	1917	Quebec, Canada
Forth (Railway)	521	1890	Queensferry, Scotland
Delaware River	501	1971	Chester, PA-Bridgeport, NJ
Greater	480	1958	New Orleans, LA
Howrah	457	1943	Calcutta, India
Transbay	427	1936	San Francisco, CA
Baton Rouge	376	1968	Baton Rouge, LA
Tappan Zee	369	1955	Tarrytown, NY
Longview	366	1930	Columbia River, WA
Queensboro	360	1909	New York, NY

TABLE 13-3
The Leading Steel Arch Bridges

Bridge	Main span (meters)	Completion date	Location
Bayonne (Kill Van Kull)	504	1931	New York, NY
Sydney Harbor	503	1932	Sydney, Australia
Port Mann	366	1964	British Columbia, Canada
Thatcher	344	1962	Panama Canal, Panama
Trois-Riveres	355	1967	St. Lawrence River, Canada
Zdakov	330	1967	Vitava River, Czechoslovakia
Runcorn-Widnes	330	1961	Marsey River, England
Birchenough	329	1935	Sabi River, S. Rhodesia
Glen Canyon	313	1959	Colorado River, AZ
Lewiston-Queenston	305	1962	Niagara River, U.S.A.–Canada

TABLE 13-4
The Leading Continuous Truss Bridges

Bridge	Main span (meters)	Completion date	Location
Astoria	376	1966	Columbia River, OR
Tenmon	300	1966	Kumamoto, Japan
Dubuque	258	1943	Mississippi River, IA
Somerset	256	1964	Tauton River, MA
Brent Spence	253	1963	Ohio River, OH
Earle S. Clements	251	1956	Shawneetown, IL
St. Louis	245	1944	Mississippi River, MO
Kingston-Rhinecliff	244	1957	Hudson River, NY
Sherman Minton	244	1961	New Albany, IN

TABLE 13-5
The Leading Cable-Stayed Bridges

Bridge	Main span (meters)	Completion date	Location
Duisburg-Neuenkamp	350	1970	Duisburg, Germany
Masopotamia	340	1972	Corrientes, Argentina
Lower Yarra	336	1970	Melborne, Australia
"Kniebrucke"	320	1969	Dusseldorf, Germany
Erskine	305	1970	Clyde River, Scotland
Bratislava	303	1971	Danube River, Czechoslovakia
Severin	302	1959	Cologne, Germany
Mannheim-Nord	290	1969	Mannheim, Germany
Wadi Kuf[a]	286	1969	Libya, Africa
Leverkusen	280	1965	Leverkusen, Germany
Friedrich Ebert	280	1967	Bonn, Germany
Ganter[a]	174	1982	Valais, Switzerland

[a]   A Concrete Bridge

TABLE 13-6
The Leading Concrete Arch Bridges

Bridge	Main span (meters)	Completion date	Location
Gladesville	305	1964	Sydney, Australia
Foz do Iguassu	290	1964	Parana River, Brazil-Paraguay
Arrabida	270	1963	Porto, Portugal
Sando	264	1943	Angerman River, Sweden
Shibenik	246	1967	Krka River, Yugoslavia
Fiumerella	231	1961	Catanzaro, Italy
Novi Sad	211	1967	Bregenz, Austria
Van Stadens Gorge	198	1970	Port Elizabeth, S. Africa
Matrin Gil	192	1942	Andavias, Spain

TABLE 13-7
Continuous Plate and Box Girder Bridges

Bridge	Main span (meters)	Completion date	Location
Sava I	75-261-75	1956	Belgrade, Yugoslavia
Zoobrucke	74-259-145	1966	Cologne, Germany
Sava II	42-250-40	1970	Belgrade, Yugoslavia
Bonn-Sud	125-230-125	1971	Bonn, Germany
San Mateo–Hayward	114-229-114	1967	San Francisco, CA
Dusseldorf-Nuess	103-206-103	1951	Germany
Schierstein	85-205-85	1962	Wiesbaden, Germany
Weisenau	74-204-134	1961	Mainz, Germany
San Diego–Coronado	201-201-171	1963	San Diego, CA
Europe	108-198-108	1963	Still River, Austria

NOTES

1. **Stranges, A. N.** Synthetic petroleum from high pressure coal hydrogenation. *Chemistry and Modern Society.* American Chemical Society (1983), p. 22.
2. Ibid.
3. **Duckert, J. M.** *A Short Energy History of the United States.* The Edison Electric Institute (1980).
4. Ibid.
5. **Chiles, J. R.** Spindle top. *American Heritage of Invention and Technology.* Vol. 3, No. 1 (1987), p. 35.
6. Ibid.
7. Ibid., p. 40.
8. **Doscher, T. M.** Enhanced recovery of oil. *Am. Sci.* **69**(2): 193 (1981).
9. *Oil Industry* Vol. 12, No. 12 (1977), p. 43.
10. Ibid., p. 58.
11. **Hambly, E. C.** The North Sea Challenge. *Great Engineers.* Walker, D., Ed. Academy-St. Martins (1987), p. 170.
12. **Robinson, R.** BULLWINKLE. *Civil Eng.* **59**(7): 34-37 (1989).
13. **Hambly,** op. cit., p. 174.
14. **Harrison, G. R.** Production in the Beaufort Sea—feasible by the mid-1980s. *Ocean Industry,* Vol. 15, No. 8 (1980), pp. 35-49.
15. **Kemp, P.** *The History of Ships.* Longmeadow (1988), pp. 265–266.
16. **Butler, J. N.** The largest oil spills. *Ocean Industry,* Vol. 13, No. 10 (1978), pp. 101–114.
17. **Skinner, S. K. and Reilly, W. K.** The EXXON VALDEZ Oil Spill. *Oil Spill Intelligence Report.* May (1989).
18. **Doscher,** op. cit., p. 194.
19. **Blotter, P. T.** *Introduction to Engineering.* John Wiley & Sons (1981), p. 75.
20. **Doscher,** op. cit., p. 194.
21. **Berg. R.** *Reservoir Sandstones.* Prentice-Hall (1987).
22. **Billington, D.** Bridges as structural art. *Blueprints* (fall) (1981), p. 10.
23. **Finch,** op. cit., p. 485.
24. **McKay, E. A.,** Tunneling to New York. *American Heritage of Inventions and Technology,* Vol. 4, No. 2 (1988), pp. 22–31.
25. **B. F. Sturtevant Company.** *The Eighth Wonder.* University Press (1927).
26. Ibid., p. 45.
27. Ibid.
28. **McKay,** op. cit., p. 31.
29. Ibid., p. 29.
30. Ibid.
31. Ibid.
32. **B. F. Sturtevant Company,** op. cit., p. 53.
33. **Flink, J. J.** The path of least resistance. *American Heritage of Innovation and Technology* Vol. 5. No. 2 (1989), p. 34.

34. **Flink, J. J.** Innovation in automotive technology. *Am. Sci.* **73**(2): 151–161 (1985).
35. **Cowan, R. S.** Less work for mother? *American Heritage of Invention and Technology,* Vol. 2, No. 3 (1987).
36. **Mitchell, J., Ed.** Railroads of the future. *Random House Encyclopedia.* (1983), pp. 1704–05.
37. **Kirby, et al.** *Engineering in History.* McGraw-Hill (1956), p. 377.
38. **Coel, M.** A silver streak. *American Heritage of Invention and Technology,* Vol. 2, No. 2 (1986), p. 10.
39. Ibid., p. 13.
40. **Mitchell,** op. cit., pp. 1706–09.
41. **Bobrick, B.** *Labyrinths of Iron.* William Morrow (1986), p. 178.
42. **Overskei, D.** Advanced Tokamaks and a path to ignition. *Physics Today,* Vol. 41, No. 1 (1988), pp. 63-65.
43. **Schwarzchild, B.** Princeton Tokamak reaches record plasma ion temperature. *Physics Today,* Vol. 39, No. 11 (1986).
44. **Bodner, S. E.** Recent progress in laser fusion. *Physics Today,* Vol. 41, No. 1 (1988), pp. 65-66.
45. **Emme, E. M., Ed.,** Perspectives on the history of flight in America. *Two Hundred Years of Flight in America.* AAA History Series, Vol. 1 (1977), p. 24.
46. Ibid.
47. **Vladimirov, L.** *The Russian Space Bluff.* London (1971).
48. **Bryan, C. D. B.** Apollo to the Moon. *The National Air and Space Museum.* H. N. Abrams (1984), p. 377.
49. **Winter, F. H.** In *The Rocket Societies: 1924-1940.* Smithsonian (1983). This work points out parallels in Korolev and Von Braun. They began as idealists where space was concerned; both were diverted by militaristic regimes and had post-war leadership of space programs.
50. *NASA, The First 25 Years, 1958-1983.* National Aeronautics and Space Administration (1983).
51. **Nicks, O. N.** *Far Travelers.* NASA (1985), p. 62.
52. Ibid., p. 63.
53. Ibid., p. 159.
54. **McBride, J. F.** Solid-propellant rocket systems. *Exploring in Aerospace Rocketry.* NASA (1971), p. 97.
55. **Nicks,** op. cit.
56. **NASA,** op. cit., pp. 95-102.
57. **Ezell, E. C.** The heroic era of manned space flight. *Two Hundred Years of Flight in America.* AAA History Series, Vol. 1 (1977), p. 241.
58. *Space Shuttle.* NASA (1975).
59. **Billington,** op. cit.
60. **Sandström, G. E.** *Man the Builder.* McGraw-Hill (1970), p. 234.
61. **Billington, D. P.** Structural art and Robert Maillart. *Civil Engineering: History, Heritage, and the Humanities (II).* Princeton (1973), p. 165.
62. **Straub, H.** *A History of Civil Engineering.* MIT Press, Cambridge, MA (1964), p. 238.
63. Ibid., p. 208.
64. Ibid., p. 210.

14 NEW TECHNOLOGY AND THE FUTURE

COMPUTERS

To date, the electronic digital computer represents man's most complex device.[1] He has never before or since produced a device where the probability of failure had to be so low (1 in 10^{14} for 12 h operation time), unless it is space capsules and shuttles with all their attendant computers. The development of the computer traces back, historically, to the first use of mechanical devices or "engine" for mathematical calculations. Gottfried Leibnitz developed a "Leibnitz Wheel", in 1673 which performed the four basic operations of addition, subtraction, multiplication, and division. Blaise Pascal had earlier (1642–44) built a machine that added and subtracted. Leibnitz's device performed the operation of multiplication automatically by repeated additions; his theme was to free men from the slavery of dull but simple tasks. Charles Babbage (1791–1871) continued Leibnitz's theme in constructing an automatic digital computing machine.

Babbage's "Difference Engine" was not largely utilized because the technology and engineering of his age was not adequate for the tolerances needed in the machine's gears and wheels.[2] By 1833 Babbage had given up on his machine. In essence, Babbage's device was used to calculate polynomial equations up to degree 6. It was not a general purpose machine but at this time the production of tables for others' use was a valid task for many mathematicians and the like.

Babbage constructed an "Analytical Engine" which was general in nature. It operated with two sets of cards, one that prescribed or set the machine for the particular formula and the second set with the specific variables to be operated upon. The Analytical Engine was thus of two parts—first, a "store" and second, a "mill". Here with just a little imagination we can see early analogues of the modern computer, e.g., the concept of a storage unit (memory) and an operating unit (control/program).

Analog vs. Digital Machines

Leibnitz's, Pascal's, and Babbage's machines are generically classified as digital or arithmetical. Digital refers to the quantities employed and arithmetical the process performed. The simplest form of digital computer is the abacus. Analog machines are very different. They utilize continuous quantities and perform measurement processes. Analog requires representation of numbers as physical qualities such as rod length, DC voltage, light, etc. A slide rule is a good example of an analog machine, i.e., the comparison of two lengths to achieve the values of mathematical operations by measurement.

A planimeter, familiar to engineers, is a device for measuring the area bounded by a simple closed curve described by the function

$$y = f(x)$$

or the equation

$$y = \int_a^b f(x)dx$$

which requires integration of the function by the analog device. This example represents a simple, first-order differential equation. Most physical behavior is described by second-order differential equations of the form

$$\frac{d}{dx}\left(\frac{1}{p}\frac{du}{dx}\right) = u$$

where p is a function of x. By a two-step integration of

$$g_1(x) = \int u_1\,dx$$

and

$$u_2(x) = \int p(x)g_1(x)dx$$

it is possible to solve this equation. The second integration result, $u_2(x)$, can be used in the first integral for u, and one integration will solve the problem. This assumes the elimination of $g_1(x)$ in the differentials where

$$dg_1 = udx, \quad du = pg_1\,dx$$

Thus, the solution

$$\frac{d}{dx}\left(\frac{1}{p}\frac{du}{dx}\right) = u$$

The particular example is based on Sir William Thomson's (Lord Kelvin) description of an integrator evaluating harmonic motion. A more arcane problem, the U.S. Census spurred computer development more than Lord Kelvin's automated evaluation of physical motion.

In 1890 it became obvious to Herman Hollerith (1860–1929) and John Shaw Billings (1839–1913) that a system of machines was necessary to adequately take the U.S. Census. Hollerith was an engineer with a doctoral degree from Columbia University. He later instructed at MIT in mechanical engineering. Hollerith patented a tabulating machine in 1889. Hollerith later associated with Thomas J. Watson, Sr., in 1914, in the Computer Tabulating-Recording Company which later (1924) became the International Business Machines Corporation (IBM). Billings was in charge of the work on vital statistics in both the 1880 and 1890 censuses. Hollerith was a member of his staff. Proceeding on Billing's suggestions, Hollerith designed a punch-card reading tabulating machine which characterized vital statistics on the cards. The roll of cards ran under a set of contact brushes which completed an electrical circuit if and only if a hole was present. The completed circuits activated counters which advanced a unit/hole counted. With this device the 1890 Census handled the records of 63,000,000 people. Even with this initial success, analog devices were those most favored by scientists of the caliber of Michelson and Kelvin. Mechanical improvements allowed successful application of two-step integrations and held sway until 1945. The development of the first successful electronic digital machine—ENIAC (1945)—and an earlier innovative design, ABC, matched closely the theoretical work of John von Neumann. The basic structure of the modern automatic computer was outlined by A.W. Burks, H.H. Goldstine, and John von Neumann in a 1946 paper.[3] Their machine required: (1) a control unit for executing orders, (2) a memory unit for storing orders and data, (3) an arithmetic-logic unit to perform elementary operations, and (4) input and output units for either machine-machine communication. This design has been termed von Neumann (computer) architec-

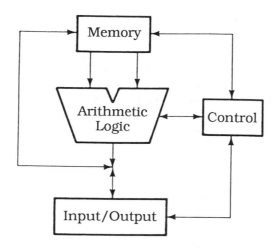

FIGURE 14-1.   Principal flow of data, instructions or
control signals within a von Neumann computer.

ture.[4] This is shown in the diagram of Figure 14-1. The designation also implies a sequence
to the processing of orders and operations termed "linear" or more commonly "serial".

The ABC computer was a pre-World War II design by John Vincent Atanasoff (1903-)
and Clifford E. Barry (hence the name "A", Atanasoff, "B", Barry, "C", computer). It
was a digital design, using binary coding and a von Neumann-like architecture.[5] Atanasoff,
electrical engineer and physicist, was concerned as Babbage and Hollerith before him, with
reducing the laborious computations in using electromechanical means. Atanasoff and his
student Barry developed a prototype by October 1939. This machine had two memories,
each capable of holding a 25-digit binary number, equivalent to an 8-digit decimal number
(for discussion of binary numbers, see below). It could only add and subtract. Building on
their initial success, they constructed the ABC between 1939 and 1942.[6]

The ABC used IBM punch cards for input. The data was converted to base 2 by a
rotating drum and stored to memory (up to 30 numbers). By elimination of constants between
the keyboard drum and counter drum the ABC could solve linear equations with up to 29
unknowns. The ABC could not reach its full potential due to reading errors every 10^4–10^5
repetitions. The device could readily solve small systems of equations. Development of the
device was halted during World War II. Flawed but demonstrably a functioning computer,
it deserves credit along with more successful contemporaries such as Turing's COLLOSUS
and Mauchley and Eckert's ENIAC.

Digital Triumph

The final arbitrar of the vicissitudes of analog and digital computing lay in the simple
notion of speed. The use of electromechanical relays required 1 to 10 ms (10^{-3} s) to open
or close, due to the inertia of the mechanical parts. In a tube or transistor the elements being
moved are electrons. These masses are 9×10^{-28} grams as compared to relay contacts
whose masses are about 1 gram. With almost no inertia to overcome, the electronic circuit
operates almost instantaneously. Total time in the early computers, like ENIAC (Electronic
Numerical Integrator and Calculator), to actuate the other parts of a circuit was 5 ms (10^{-6}
sec). At the present time electronic circuits operate in nanoseconds (10^{-9} s). ENIAC,
developed by John W. Mauchley and J. Prosper Eckert at the University of Pennsylvania,
School of Electrical Engineering, had microsecond speed that simply was a thousand times
faster than the best analog relay machines. Why was this speed so important? Simply because
many physical problems involve multiple parameters such as weather with say n multipli-

cations at t time intervals over 5 space points. Then the total calculation is proportional to $nt(1)$, nts, nts^2, nts^3,..., nts^n. Thus, for three spatial dimensions one can quickly have 750 million multiplications.

Another factor involved in the need for such computing speed is the use of a binary number system. In the binary system there are only two kinds of digits—0 and 1—instead of 10 (1–10) in standard digits. Addition of binary digits is simply: $0 + 0 = 0 + 1, 1 + 0 = 1. 1 + 1 = 10$ (this merely says $1 + 1 = 2$, which is represented in the binary form as $1 \times 2^1 + 0$ or in shorthand as 10, "one zero"). Likewise, multiplication uses only the four forms:

$$0 \times 0 = 0, 1 \times 0 = 0, 0 \times 1 = 0, 1 \times 1 = 1.$$

For a number like 141, it is written binarily as 1000 1101 or

$$128 + 8 + 4 + 1 = 1 \times 2^7 + 0 \times 2^5 + 0 \times 2^4 + 1 \times 2^3 + 1 \times 2^2 + 0 \times 2^1 + 1.$$

In this system a Binary digit is a "bit". Eight bits equal a byte (which corresponds to the number of data pins and lines of a data bus). One half a byte is a nibble ("nybble"). Typically, one byte binary values are expressed in sets of nibbles like our number 141 is more correctly written 1000 1101.[7] Inherent in the use of the binary system in computers is the obvious increase in steps necessary to represent numbers in binary form. For instance, to multiply two ten decimal digits requires 1—individual products plus additions. The same numbers require 32 binary digits to represent them and 1000 individual steps. In the economy of speed, the tenfold increase in the number of computing steps is offset by the 1000-fold decrease in time necessary to accomplish them. Herein lay the triumph of the electronic digital computer. An interesting footnote to this discussion is the mode of operation for the computer. With the first machines like ENIAC (Table 14-1), the computer operated in a serial mode as compared to parallel mode. In serial mode in an operation like addition all corresponding pairs of digits are added serially in time. The difference lies with the parallel mode of operation adding corresponding pairs of digits simultaneously. Ultimately, the parallel mode of operation on binary numbers prevailed in modern computers (Figure 14-2). The serial mode has been maintained in modern machines such as microcomputers in that one instruction at a time is inspected and executed by the device.

BIOMEDICAL ENGINEERING

Biomedical engineering is one of the newest areas of modern engineering and describes the activity at the interface of medicine and engineering. Its most singular accomplishment to date has been the total substitution of a mechanical heart for a faulty organic one.[8] While a stunning accomplishment both from a medical and engineering sense, the substitution of a mechanical device for an organic device is less dramatic when one considers the cardio-vascular system as a closed hydraulic system with a four-chamber pump.

It is this dispassionate characterization of physiological process into hydraulic (heart, circulatory system), pneumatic (respiratory system), and communication (nervous system) systems that allows the biomedical engineer to design monitoring and control systems for the human "hemodynamic system" and other organs.

One of the first attempts to monitor body functions involved the recognition of bioelectric potentials associated with nervous activity, heartbeat, muscular activity, etc. These potentials are ionic voltages produced by cellular electrochemical activity. Luigi Galvani, in 1786, recognized that electricity was generated by the body. To measure these potentials a trans-ducer capable of converting ionic potential and currents into electric counterparts is necessary. The first of these developed were simple, two-pronged electrodes. These electrodes recorded

TABLE 14-1
Engineering Advances in Computers

1945	ENIAC calculates 5000 additions/second; 500 multiplications/second; all tube construction
1947	Ferrite memory suggested by J.W. Forrester*
1951	Parallel or RAM memory developed
1951	CRT display
1955	IBM introduces card programming
1945–1956	FORTRAN programming language developed
1959	Solid state machines replace early tube-type machines. Transistors had smaller size, reduced power requirements and heat generation, and greater reliability. The transistor is a silicon crystal "doped" (has added to it) phosphorus and boron to provide a p-n semiconductor junction. This junction allows amplification (gain) of a base current applied to the p-side of the junction whereby excess charges in the n-types semiconductor flow out the transistor. Memory access was reduced to microseconds. The first commercially available computer was the UNIVAC I.
1965	Integrated circuits, with thousands of transistors fabricated on the surface of a silicon chip less than the size of postage stamp, replace transistors. Memory access time reduced to nanoseconds. Up to 65,000 components on a single chip. Magnetic disc storage.
1970s	VLSI or "very large scale integration" denotes the increase of circuit densities of 100,000 components per chip. Memory access time approaching 1 ns.
1980s	VLSI continues with circuit densities to increase of 100 over current chips. New memory devices such as tunnel junctions that operate at temperatures near absolute zero. These memories will have memory access time expressed in picoseconds (trillionths of a second). The basis of the tunnel junction lies in quantum theory wherein electrons can penetrate energy barriers they ordinarily cannot surmount. By cooling to $-269°C$ the superconducting state is achieved with almost frictionless transmission of current.

* Forrester, J.W. Data Storage in Three Dimensions. Project Whirlwind Report No. 70. Massachusetts Institute of Technology, Cambridge, MA (1947).

characteristic wave forms associated with electric fields generated by the ionic currents. Examples of these are those of the EKG and EEG, electrocardiogram and electroencephalogram, respectively.

The electrocardiogram's earliest bioelectric potentials were made with immersion electrodes—buckets of salt water in which the subject placed one hand and one foot. The principle of the EKG was developed by Willem Einthoven in 1903. An EKG was built in 1912 which embodied most of the functions familiar in today's machines.

The EEG measures neuronal activity of the brain. Originally it was believed that the EEG potentials represented a summation of potentials of neurons in the brain. Today's theories place the source of the EEG patterns at the synaptic junction of the neurons. The correlation of EEG patterns with specific brain activity is only a relatively general level at present—delta, theta, alpha, and beta ryhthms. For instance, alpha range patterns (8 to 13 Hz) correspond to relaxation states or a resting individual. Alert states tend to demonstrate an unsynchronized high-frequency pattern in the EEG.

Other types of bioelectric potential devices include the EMG (electromyogram), the EGG (electroretinogram) and EOG (electrooculogram). These latter devices are even less precise in intercorrelation of patterns to activities of the eye and muscles. It is in the area of electrode measurement of bioelectric potentials that biomedical engineering made its first significant contributions to medicine. Modern intensive care units utilize variations on these type monitoring systems with additional instrumental observations of respirations and temperature.

Monitoring of skin temperature by infrared devices has led to the development of the thermograph—a scanning device that maps the infrared energy onto photographic paper producing a thermogram. The mapped variations in systematic temperature, over the body, can determine areas of anomalous thermal activity, i.e., warm areas correlate with some tumors and vascular blockages produce cooler areas. In this configuration temperature mea-

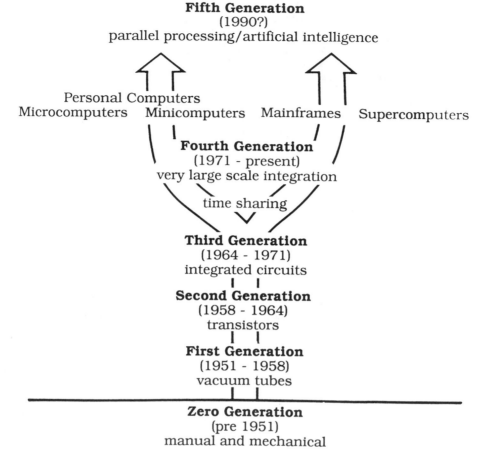

FIGURE 14-2.   The evolution of the modern computer era can be conceptualized as a succession of generations closely linked to technological advances. In the third generation, a divergence emerged with the advent of machines expressly designed for the needs of individuals. Today, a broad range of computers are available. The progression seems to be moving toward a fifth generation in which artificial intelligence will figure prominently. (Redrawn from Chapra and Canale, 1985.)

surement is done for diagnostic purposes rather than just monitoring purposes. Indeed, almost any of the monitoring devices developed by biomedical engineers can and are used extensively in diagnosis.

One recent interesting development in monitoring systems has been the extensive use of biotelemetry in the exploration of space. Biotelemetry is the measurement of biological parameters at a distance. Einthoven, the inventor of the electrocardiogram, transmitted electrocardiograms from a hospital to his laboratory many miles away as early as 1903.[9] His method was by "hard wire" transmission of biodata. Modern telemetry has eliminated the use of wires, substituting some form of modulation of a radio-frequency carrier (AM or FM) and is more properly known as radio telemetry. Parameters measurable by biotelemetry by 1963, in NASA's Mercury Program included:

1. Temperature by thermistor
2. Respiration by impedance pneumograph
3. Electrocardiograms
4. Indirect blood pressure by contact microphone and cuff

With the impetus from NASA, modern biotelemetry has expanded to include almost any biological parameter to include EKG, EMG, and EEG, as well as any physiological variable (blood pressure, gastrointestinal pressure, blood flow, and temperature). Typically, the physiological parameter is measured or converted in electrical form. The appropriate analog signal (because it is continuous measurement) is converted into a code for transmission on the carrier frequency. After reception the code is converted back to electrical form and displaced on the appropriate monitoring instrumentation, i.e., oscilloscope, rate-meter, etc.

The code is impressed on the carrier by modulation of a transmitter. Demodulation by the receiver deciphers the information from the carrier. The most common modulation system in telemetry is frequency modulation (FM) where the frequency of the carrier is caused to vary with the modulated signal. FM is much less susceptible to interference because variations in amplitude of the received signal caused by interference can be removed at the receiver before demodulations. In biotelemetry the physiological signal is sometimes used to modulate a low-frequency carrier, or subcarrier, often in the audio-frequency range. When transmitting several physiological signals, each is placed on a subcarrier of a different frequency and then all are combined to simultaneously modulate the RF carrier. The multichannel technique is termed frequency multiplexing and is economical in that a multiplexed signal can use only one transmitter and receiver for the various subcarriers.

The other approach in biotelemetry is pulse modulation of the RF carrier into a series of short bursts or pulses. Amplitude or width of pulses can be varied to carry the physiological information. To multiplex a pulsed modulation carrier, time duration is used, or in other words, for six subcarriers, the series of six pulses is composed of one pulse/signal, which amplitude or width yields the transmitted information. As we have noted, NASA was a prime force in the development of biotelemetry which today benefits both out-patient and emergency patient monitoring away from permanent treatment facilities.

Diagnostic, Control and Treatment Systems

In these two areas, biomedical engineering has made its most dramatic advances. Under diagnostic devices we have:

1. X-Ray and radioisotopic systems
2. Ultrasonic Systems
3. Particle Beam Accelerators

Under control systems we have the development of:

1. Pacemakers
2. Defibrillator
3. Artificial organs

Radiology

X-rays were discovered by Conrad Roentgen in 1895. This penetrating radiation was quickly recognized for its importance to medical diagnosis. Later it was utilized for therapy. X-ray machines were the first widely used electrical machines in medicine. X-rays are generated when fast moving electrons suddenly decelerate when striking a target. X-ray intensity varies with current. With wave-lengths in the 10^{-6} cm range they have little difficulty passing through most matter. The attenuation of a few or several of the impinging X-rays is reflected in densities of activated photoemulsion on an X-ray film. The discrimination of X-rays is maximized in the dense tissues such as bone while most body organs are relatively transparent to X-rays. These latter areas can be viewed by using contrast media prior to taking the X-ray. Air is used as a contrast medium in examining the brain, while barium sulfate gives greater contrast to the gastrointestinal tract.

The basic limitation of X-ray images is that they are two-dimensional representations of three-dimensional structures. The presence of an organ in front or behind the one of interest obscures the detail in the X-ray image. To overcome this, stereoradiography and tomography have been developed. The latter machines scan body sections with a thin X-ray beam and determine X-ray absorption with a radiation detector. An image of the section is reconstructed from a large number of scans with the aid of a digital computer.

Radioisotopes

While X-rays are radiation actively generated by devices, radioisotopic techniques rely on the inherent capacity of radioactive elements such as Phosphorus-34, Iodine-131 and Molybdenum-99 to generate emissions detectable by counting or scanning devices. By introducing these elements, typically in solution, specific areas of the body can be evaluated for physical variations in blood flow, organs, and physiological processes of extra- or intracellular respiration. These radioisotopes or elements can be counted *in vitro* in or outside the body.

The radioelements utilized tend to have short decay periods or "half lives" of minutes or a few hours. This is important because of the deleterious effect of this ionizing radiation. All natural radiation, X-ray, gamma, alpha, or beta, causes ionization of surrounding tissues. The amount of tissue damage is directly proportional to the kind and amount of the specific radiation type, with alpha particles (helium nuclei) being the most damaging to biological tissue. This component of radiation effect has a useful role and has been utilized in the therapy of tumors. This form of radioisotope utilization is broadly termed radiation therapy.

In radiation therapy the ionizing effect of the various types of radiation is used in the treatment of tumors within the body. The use of radioactive gold (Au^{198}) in prostate cancer has been developed to use the implantation of small pellets into the gland's area. Gamma rays (very high energy X-rays) continually bathe the tumorous area, with destruction of the diseased cells. The pellets utilized in this procedure are prepared in neutron irradiation of nonradioactive gold in reactors.

Gamma ray sources, such as Cobalt-60, have been engineered into devices that completely circumscribe areas of the patient, bombarding diseased areas. Even more advanced engineering is involved in the increasing use of linear accelerators to produce high energy X-rays for tumor therapy. Solid targets of various materials are placed in beams of neutrons, electrons, or mesons that produce high energy, penetrating radiation of very high flux (cross-sectional intensity). This high density of radiation reduces total patient exposure while maximizing that of the tumor.

Ultrasound

Ultrasound is sonic energy above 20 kHz. Its use dates back to the period immediately following World War II and is an outgrowth of the military development of SONAR. Ultrasound penetrates a media like the body and part of the beam is reflected and the remainder refracted. Using characteristic impedance and attenuation values for various body tissues, ultrasound can be used for noninvasive imaging of internal structures.

Various scan modes are used in ultrasound imaging and these are, A-scan, M-scan, and B-scan. A-scan displays pulse return heights on an oscilloscope. M-scan creates a record of point echo returns on the oscilloscope. Both A- and M-scan modes use a stationary transducer recording movement of interior organs and structures. The M-scan of the heart, when recorded on strip chart paper, reminds one of a sonar record recording movement of echoes with respect to time.

B-scan presents a two-dimensional image of a stationary organ or structure. The B-scan transducer is moved with respect to the body while the vertical displacement of the return pulse on the oscilloscope or chart is made to correspond to the movement of the transducer. B-scans are the most popular in diagnosis today, with extensive usage made of the technique in pediatrics. Because of the short wavelengths used in ultrasound, resolution of small internal structures is excellent. Typical frequencies used include: 2.25 MHz for general purpose studies, 3.5 MHz for studies of children, and 5.0 MHz for pediatric echo cardiography. Typical pulse rate scans are 1000/s.

Pacemakers and the Defibrillator

These control devices of cardiac activity are two of the most successful developments of biomedical engineering. Both devices artificially generate control stimuli for heart action. The differences lie in the duration and the amount of stimulus rather than kind.

Pacemakers are devices which generate artifical pacing impulses and deliver them to the heart. When normal heart electrical activity is blocked by heart damage due to disease, it is possible to supply a corrective signal by external means such as implanted electrodes.

Two varieties of pacemakers have come into usage since the 1960s—internal (implanted) and external. Internal pacemakers are implanted with the pulse generator placed in a surgically formed pocket below the clavicle (collar bone), the left subcostal area or in women, below the pectoralis major muscle. Electrodes, with leads, are attached to the myocardium. External pacemakers are used on patients with temporary heart irregularities. The pulse generator is worn on the belt or wrist, with the electrodes placed over the heart area.

The pacemaker is classified as either competitive or noncompetitive. The noncompetitive device is synchronized with electrical activity of either the atria or the ventricle. When all or part of the heart's electrical conduction is irregular, the pacemaker functions as a demand system, completely controlling the heart rate. This latter mode describes the competitive pacemaker configuration which characterized all the early devices. The competitive pacemaker operates on a fixed rate, where the noncompetitive pacemaker can adjust to physiological changes.

The defibrillator is used when the heart's activity becomes wholly unsynchronized; a condition results which is termed fibrillation. It is characterized by rapid irregular contractions of the myocardium. It can occur in the atria or ventricles. Venticular fibrillation is the most dangerous, with death occurring in minutes. The most successful development in countering fibrillation is the defibrillator. The first devices applied a brief (0.25 – 1 s) burst of 60 Hz AC of 6 A countershock which resynchronizes the heart. This technique is termed AC defibrillation.

In 1962 Bernard Lown of Harvard developed a new method termed DC defibrillation. In this method a capacitor is charged to high voltage and rapidly discharged through electrodes across the chest of the patient. This defibrillation method was found to be more successful than the AC method in correcting ventricular fibrillation. The amount of energy applied to the heart varies between 100-400 Watt second or joules. The discharge is for 5 μs. Because of potential damage to the myocardium caused by voltages as high as 6000 V some new defibrillators use longer duration discharge (10 ms) at much lower voltages. Newer models can be synchronized with the appropriate part of the EKG wave form or pulse and functions much like a very powerful pacemaker.

Artificial Organs

The decades of the 1960s and 1980s have seen the most dramatic biomedical engineering creations in the form of artifical devices that can substitute for natural organs performing life-sustaining processes. The three most prominent devices of this type are:

1. The artificial kidney or dialysis machine
2. The heart-lung machine
3. The artificial heart

Artificial Kidney

The first dialyzer used as an artificial kidney was built by Able, Rowntree and Turner in 1913.[10] The concept of dialysis is the minute rate of net exchange of a substance per unit concentration gradient between plasma and dialyzing fluid. Hemodialysis refers to the removal of undesirable molecules from blood through a semipermeable membrane into a dialyzing bath. Artificial kidneys are, then, hemodialyzers.

The first successful artificial kidney was designed by Kloff in 1940 in Holland. It used a rotating coil, where blood was pumped through cellophane tubing round about a drum. The drum is bathed in osmotic volute to remove water and associated impurities. Further improvements have involved parallel flow arrangement allowing ultrafiltration by increased hydrostatic pressure of blood on the filter membrane. The basic advantage of the latter method lies in the reduced amount of blood necessary to prime the dialyzer ($<$500 ml). The emphasis in today's developments is to reduce the size of dialyzer and to increase simplicity. This can yield home dialysis with less disruption of a patient's daily routine than with today's hospital visit regimes.

Extracorporeal Circulation Devices

The first successful substitute for the heart and lungs, used in cardiac surgery, was developed by J.H. Gibbon in 1953.[11] This successful machine was developed with the aid of the engineering staff of IBM.[12] Successfully applying principles of hydraulics, together with understanding of critical physiological parameters, allowed this team to circulate and oxygenate blood outside the body for several hours with no ill effect to the patient.

As blood is considered a tissue, circulation of it through pumps and tubing can cause trauma to its cellular makeup. To prevent this, heparin is used to prevent clotting of any injured cells. Heat exchanges prevent the blood temperature from falling below body-ambient level. Hemolysis must be prevented and the venous flow must be saturated with oxygen while removing the carbon dioxide. The oxygenator (lung) must not traumatize the blood. All these were conditions in Gibbon's design. The blood was moved by roller- and finger-type pumps. These pumps have no valves and have no direct contact with the blood. They have constant pressure at flows up to 6 liters per minute. The finger-type pump uses the principle of serial compression of an elastic tube by parallel metal arms. In the roller-type pump, two or more rollers, mounted at the ends of a horizontal revolving bar, perform successive compression of tubing. As one roller leaves the tube, the second begins to compress it. Re-expansion of the tubes after compression provides the required suction of the pump. A typical anesthetized patient has a cardiac output of 3 to 4 liters per minute per square meter of body area. Flows 2.4 l/min do not produce metabolic difficulties. Body temperature plays a role here in that below 30°C, flow rates should be reduced as oxygen requirements are reduced by half.

The artificial lung must oxygenate at least 5 liters of blood at no less than 95% saturation. Carbon dioxide must be removed at normal CO_2 tension for the arterial blood. Three types of artificial lungs have been utilized:

1.   Screen oxygenators
2.   Disc oxygenators
3.   Bubble oxygenators

The screen oxygenator was used by Gibbon and produces the least alteration in the blood tissue. Due to the differential laminellar flow rates across the gas exchange screen's surface, those faster flowing layers away from the screen have lower efficiencies. Gibbon's screens were stainless steel of 0.7 mm diameter mesh.

Disc oxygenators are revolving wheels that dip into the blood flow, performing the same function as the screen version. High rotation rates can produce foaming, with bubble uptake by the blood posing embolism problems. The same problem exists for the bubble oxygenator and the vertical screen, with different materials for the mesh being tried for gas exchange, synthetic fiber being the most common type in use today.

Artificial Heart

The heart-lung machine obviously performs the function of the human heart. The difference between it and the artificial heart pump is the duration of the cardiac assistance provided by the two types of devices.

The first implantation of an artificial component of the circulatory system was done by Hufnagel, in 1951,[13] when he placed a ball check-valve in the descending aorta. Clinical success was tempered in subsequent cases by failures due to embolism distal to the valve, rupture of the vessel at valve junctions and improper action relative to aortic regurgitation. Leaflet valves were tried in the late 1950s to replace the critical aortic valve but fatigue of material, fracture and stiffening (with scar formation) negated early work. Better ball-valve designs of the 1960s has led to longer term success with artificial aortic and mitral valve. The principal failures in late post-operative stages (2 years) were generally associated with improper tissue healing around the valve implant and blood clot formation.

Experimentation with partial or whole cardiac replacement began in earnest after the beginning of heart transplant surgery by Christian Barnard in 1967.[14] Denton Cooley transplanted the first artificial device into a patient in 1981.[15] This design functioned successfully as a heart replacement until a suitable human transplant could be found.

The successful implantation of an artificial heart differs from the 1982 implant in that the ''Jarvik 7'' is intended to be a permanent substitute for the organic heart. From a clinical standpoint the principle is now a feasible alternative to organic heart transplants. The long-term success of the devices has experimental support from work in laboratory animals. It is interesting and somewhat sobering, to note that the latest artificial heart implant had a ball-valve failure which required replacement shortly after the pump was implanted. The success of these prosthetic devices are the result of advances in control miniaturization and the development of synthetic materials which are both the result of advances in 20th century engineering.

Robotics

The progress of engineered technology has reached the era of automated mechanization. Industrial engineers today and in the future will rely increasingly on control systems and automation. These are and will be robots, robots not of the type that writers like Isaac Asimov envisaged but devices with artificial intelligence performing difficult or sophisticated tasks. An engineer who read Asimov, Joseph Engleberger (1925-), together with George Devol, an inventor, formed the world's first firm to build robots.[16] Following Devol's patent the first robot was a machine that unloaded hot metal from a casting machine at a General Motors factory. By the 1970s there were around 10,000 robots in use and by 1981 there were 20,000.[17]

Industry, most importantly that of the Japanese auto industry, has wed itself to industrial automation. This is true of their watch industry as well. For ordinary repetitive tasks such as welding or for detailed fine technology like miniature assembly, the robot has dramatically contributed to quality and production. The Japanese estimate that robot production will be over 100,000 per year by 1990.[18]

In 1952, John von Neumann began a book called *The Theory of Automata Construction, Reproduction, Homogenity.* It was not finished before his death in 1957, but was ultimately published by Arthur Burks as the *Theory of Self Producing Automata.* In this book von Neumann proves with mathematical rigor the possibility of building self-reproducing machines. The simplest of these is shown in Figure 14-3, where von Neumann imagined a stockroom filled with parts such that a machine with a memory tape containing the necessary

FIGURE 14-3.   Self-replication robots. (From Advanced Automation for Space Missions, 1980.)

instructions could build a replica of itself. It would transfer a copy of its memory and the replica would then duplicate itself. One can see this process is exponential. In 10 years there are 100 machines, in 20 a million.

Replicating robots or "von Neumann machines" have been discussed seriously for use in space exploration.[19,20] It may be man will conquer space vicariously through his machines.

ENGINEERING IN THE FUTURE

The engineering of the future will be distinguished by scale—materials science and structural design will allow for the achievement of longer span bridges, greater height in building and durability in construction. Computer technology seeks the ultimate micro-circuit. Engineering practice will be matched against greater environmental challenges, both in terms of its impact on designs and the impact of design on the environment.

Most of the civil engineering of today has been in the temperate regions of the world. As more environmentally marginal lands are settled or utilized the challenge to engineering will be greater. We have already seen this in the engineering for the North Slope and the Alaskan Pipeline. Extreme climate renders designs for less harsh areas ineffective. The engineering of the sea and space will create much of the problems seen in climate variation and its effect on engineered works. The behavior of materials in a frigid vacuum or under great pressures in a corrosive environment will be a serious test of future engineering. Already we have seen the rebirth of the near-extinct science of ceramics with new and exciting applications in nuclear energy and space.

It is difficult to project what exact forms the engineering of the future will take but some ideas are possible based on the "dreams" of engineers of the past. One audacious example was that of the "Panropa Plan" ("Atlantropa Plan").[21]

In 1928 Herman Sorgel (German) proposed a plan under the title *Mittlemeer Senkung* ("Reduction of the Mediterranean"). Sorgel pointed out the Mediterranean Sea was a recent geological feature (<100,000 years). The level of the Mediterranean about 50,000 years ago was almost 1000 meters lower than it is now. At this level 44% of the sea's area was dry land, surrounding two large lakes—one east of Italy/Sicily and one west. With the end

TABLE 14-2
Water Budget for
Mediterranean per
Year

Source	Percent
Atlantic	66.65
Rain	24.11
Black Sea	3.68
Rivers	5.56
Total	**100.00**

of the glaciers the lands were drowned. All we have to do to get this land back is plug the Strait of Gibraltar with an enormous dam.

The area of the Mediterranean is 1,560,000 square kilometers (ten times the size of Wyoming). It has a high evaporation rate—1.6 meters per year before replacement, or 4144 km³ area. Four major rivers drain into the Mediterranean—Nile, Po, Rhone, and Ebro. The balance of evaporation losses come from the Atlantic and the Black Sea (Table 14-2). The flow from the Atlantic is 1 million cubic meters per second (12 times the water over Niagara Falls at high water). The Panropa Plan is based on this table—dam the Atlantic entrance, dam the Black Sea at the Dardanelles, and the sea will go down.

Sorgel's Gibraltar dam would not go across the Strait at its narrowest point because the water depth is 500+ meters. The dam would be built at a point where water depths are about 300 meters. The length of the dam would be 29 kilometers; its Crown would be 50 meters wide with a foundation 500 meters wide.

Recession in the sea level would be 100 centimeters per year or 10 meters in 10 years. This pressure head could be used for electric power generation. In 100 years the level would be down 305 meters, with 150,000 square kilometers of new or reclaimed land. The two Islands of Mallorca and Minorca would be one—Corsica and Sardinia would be one. Only narrow straits would separate Sicily from Tunisia and Italy.

Sorgel then proposed a second step—two more dams—one between Italy and Sicily and one between Tunisia and Sicily. With these dams the Western half of the sea level would be completely controlled by inflow at the first dam. The eastern half would be allowed to sink another 100 meters over the following 100 years.

The rise in all other seas would be 1 meter. In the Mediterranean area the final result would be 354,000 square kilometers of new land with hydroelectric paver of almost unlimited power at several points—Gibraltar Dam, Mouths of the Ebro, Rhone, Po and Nile, the two Sicily Dams. Sorgel considered the final goal of the plan to be the fusion of the European and African Continents.

Such a plan has ramifications of significant proportions—political, economical and environmental. One obvious environmental result would be the change in course of the Gulf Stream to the English Channel, altering Western Europe's weather and fishing. In economic terms many experts who have commented on the plan suggest the level of the sea be dropped only 20 meters. Once beyond 30 meters all major ports on the sea are no longer harbors. Some point out of the problem of vulcanism in the Mediterranean area—removal of a significant weight of water might destablize the seismicity of some areas. This is not well understood and may not be a valid objection. Of course the political situation today makes the plan impossible. No Panropa Plan or an equivalent can be given serious consideration until a world environmental plan is in place.

''Firm Power'' from the Sun—Satellite Station

The problem in utilizing solar energy is twofold—collecting and converting the energy

TABLE 14-3
Sunshine (%)

Place	Winter	Summer
Boston	51	61
New York	56	67
Miami	70	65
Chicago	44	70
Los Angeles	72	75
Seattle	29	62

is the first aspect, storing the energy is the second. The second aspect is the most difficult by today's standards. Further, for industrial needs the supply must be reliable and nearly continuous. In the U.S. the desert is the only place we can assure nearly 100% sunshine. Table 14-3 shows calculated sunshine across the U.S.

The total energy output of the sun is incredible; even the fraction intercepted by the earth every second is formidable—6,000,000 tons of coal equivalence. In terms of horse-power, theoretically, a single hectare would receive 18,000 hp per second. Heat losses to the atmosphere would reduce this to 14000 hp. Present-day collectors require adequate space between them so 2/3 of the hectare cannot be occupied—leaving an energy yield at 4500 hp. Since the collection and steam generation process is no longer 100% efficient, our final yield can be reasonably estimated at 20–34 hp/hectare. Recalling our table, we would be optimistic to expect this production much more than 50% of the time.

"Firm Power" from the sun can be found only in space. Efficiency of collection and conversion to power is greatest above our atmosphere. NASA has proposed to exploit this fact by placing giant satellite collectors in fixed orbit above the earth. These stations would convert the incoming solar energy to a transmittable form such as microwaves. This low frequency, high intensity signal could be beamed to surface stations on earth for distribution in a power grid in various areas. The problems in setting up such as system are becoming less each day with the demonstrated capability of the Shuttle, as well as using existing collector technologies. The problem of storage would be moot with the availability of "firm power" 100% of the time.

Automation, Information, and Mechanization

The 21st century will see the total mechanization of work—to a greater degree engineering will be an automated, cybernetic, mechanized process. A fact of life in technology today is the increased dissemination of available information. To process this information, the computer is the calculator of the future engineers. Robotics is profoundly altering construction and maintenance processes. The engineer may be cast more and more in roles of management in design without even seeing the product of his design. We may have an analogous problem to that of the Roman emperor Vespesian, who asked.."what will I do with my poor?..." when he was offered prime movers other than human power.

It is part of the engineer's responsibility to consider the results and impact of his work on the total environment of his world. The sociological impact of a future mechanized, automated world can only be guessed. In their designs they will confront these and other profound questions that involve the ability to create engineered works and products. Part of the reason for this book is to give the reader an historical perspective with which to measure the effect of these future works.

The old lay..."to know where you're going, it helps to know where you've been". I hope this book has helped you to understand where engineers have been.

NOTES

1. **Goldstine, H. H.** *The Computer.* Princeton Press. p. 153.
2. **Goldstine,** p. 12. Compare this to the Antikyra Device's precision workmanship in antiquity.
3. **Taub, A. H., Ed.** *Collected Works of John von Neumann,* Vol. 5, Macmillan (1963), pp. 34–79.
4. **Doty, K. L.** *Fundamentals of Microcomputer Architecture.* Matrix (1979), pp. 6–16.
5. **Makintosh, A. R.** The first electronic computer. *Physics Today,* Vol. 40, No. 5 (1987), pp.25–32.
6. Ibid. p. 28.
7. **Newell, S.** *Introduction to Microcomputing.* Harper and Row (1982) pp. 18–29.
8. The "Jarvik 7" was transplanted in December 1982, into Dr. Barney Clark, a dentist, at Salt Lake City, Utah.
9. **Cromwell, L. et al.** *Biomedical Instrumentation and Measurements.* Prentice-Hall (1980), p. 317.
10. **Abel, J., Rowntree, L. G., and Turner, B. B.** On the removal of diffusible substances from the circulating blood of living animals by dialysis. *Pharmacol. Exp. Ther.* **5**: 275 (1913).
11. **Gibbon, J. H., Jr.** Application of a mechanical heart and lung apparatus to cardiac surgery. *Minnesota Med.* p. 37, (1954).
12. **Taylor, R.** A mechanical heart-lung apparatus. *IBM J. Res. Develop.* **1**: 300 (1957).
13. **Hufnagel, C. A.** Aortic plastic valvular prothesis. *Bull. Georgetown Univ. Med. Ctr.* **4**: 28 (1951).
14. **Barnard, C. and Curtis, B.** *Christian Barnard: One Life.* Macmillan (1970).
15. *Science Digest.* February 1983.
16. **Marsh, P.** *The Robot Age.* Abacus (1982).
17. Ibid., p. 67.
18. Ibid., p. 68.
19. *Advanced Automation for Space Missions, A Technical Summary.* University of Santa Clara (1980).
20. *Extra-Terrestrial Materials Processing and Construction.* NASA (1980).
21. **Ley, W.** *Engineer's Dreams.* Viking (1954).

BIBLIOGRAPHY

Abel, J., Rowntree, L. G., and Turner, B. B. On the removal of diffusible substances from the circulating blood of living animals by dialysis, *Pharmacol. Exp. Ther.* 5, p. 275 (1913).

Abernathy, W. J. and Utterback, J. M. Patterns of industrial innovation, *Technol. Rev.* 80(7) (1978).

Advanced Automation for Space Missions, A Technical Summary. University of Santa Clara (1980).

Agricola, G. *De re metallica.* (1556). Dibner Collection, National Museum of American History, Washington.

Alberti, Leone Battista. *De re aedificatoria,* N. Lavrentic (1485). Dibner Collection, National Museum of American History, Washington.

Allen, F. The letter that changed the way we fly, *American Heritage of Invention and Technology,* 4(2), pp. 6–13 (1988).

Almgren, B. *The Viking,* Crescent (1975).

Anderson W. J. and Spiers, P. *The Architecture of Ancient Rome,* Books for Libraries Press (1927).

Ashby, T. *The Aqueducts of Ancient Rome,* McGrath (1973).

Aston, J. and Story, E. B., *Wrought Iron,* A. M. Byers (1957).

Atkinson, R. S. C. Moonshine on Stonehenge, *Antiquity,* 43, pp. 212–216 (1969).

Attenborough, D. *The First Eden,* Little, Brown (1987).

Baradez, J. "Réseau routier de la zone arrière du limes de Numide," *Limes-Studien,* II, 59, pp. 19–30, Bâle (1957).

Barnard, C. and Cutris, B. *Christian Barnard: One Life,* Macmillan (1970).

Beaver, H. E. C. The growth of public opinion, *Problems and Control of Air Pollution,* F. S. Mallette. Reinhold (1955).

Beckett, D. *Brunel's Britain,* David S. Charles (1980).

Bélidor, B. F. *Architecture Hydraulique,* Vol. IV, Paris (1752–1782).

Bélidor, B. F. *Science des Ingénieurs,* Paris (1729), in Straub (1964).

B. F. Sturtevant Company. *The Eighth Wonder,* University Press (1927).

Berg, R. *Reservoir Sandstones,* Prentice-Hall (1987).

Bill, M. Maillart and the artistic expression of concrete construction, *The Maillart Papers,* from The Second Princeton National Conference on Civil Engineering: History, Heritage and the Humanities, October 1972, Princeton (1973).

Billington, D. F. Structural art and Robert Maillart, *Civil Engineering: History, Heritage, and the Humanities,* II, Princeton (1973).

Billington, D. P. Bridges as structural art, *Blueprints,* (fall 1981).

Billington, D. P. Bridges and the new art of structural engineering, *Am. Sci.,* 72(1), p. 22 (1984).

Biringuccio, V. *De re pirotechnica,* (1540). Dibner Collection, National Museum of American History, Washington.

Biswas, A. K. *History of Hydrology.* North Holland (1972).

Blotter, P. T. *Introduction to Engineering,* John Wiley & Sons (1981).

Boas, M. "Hero's *Pneumatica,* a study of its transmission and influence." Seminar in the History of Science, Cornell (1960).

Bobrick, B. *Labyrinths of Iron.* William Morrow (1986).

Bodner, S. E. Recent progress in laser fusion, *Phys. Today,* 41(1) (1988).

Boyne, W. J. *The Aircraft Treasures of Silver Hill,* Rawson Associates (1982).

Boyne, W. J. *The Smithsonian Book of Flight,* Orion (1987).

Boure, T. *The Sketchbook of Villard de Honnecourt,* Indiana University Press (1959).

Brady, S. G. *Caesar's Gallic Campaigns,* The Military Service Publishing Co. (1947).

Brett, M. *The Moors,* Golden (1985).

Bryan, C. D. B. Apollo to the Moon, *The National Air and Space Museum,* H. N. Abrams (1984).

Bur, M. The social influence of the motte-and-bailey castle, *Sci. Am.,* 248(5), pp. 132–139 (1983).

Burke, J. *The Day the Universe Changed,* Little, Brown (1985).

Burl, A. Dating the British stone circles, *Am. Sci.,* 61(2), pp. 167–173 (1973).

Butler, J. N. The largest oil spills, *Ocean Industry,* 13(10), pp. 101–114 (1978).

Butzer, K. *Environment and Archaeology,* Aldine (1971).

Caldwell, J. B. The Great Ships, *The Works of Isambard Kingdom Brunel,* Pugsley, A., Ed., Cambridge (1976).

Carnot, N. L. S. *Reflexions sur la Puissance Motrice de Feu,* (1824).

Carpenter, R. *The Architects of the Parthenon,* Penguin (1970).

Carvill, J. *Famous Names in Engineering,* Butterworths (1981).

Casson, L. *Illustrated History of Ships and Boats,* Doubleday (1964).

Champion et al. *Prehistoric Europe,* Academic (1984).

Charles, J. A. The development of the usage of tin and tin bronze: some problems, in *The Search for Ancient Tin,* Franklin, A. D., Olin, J. S., and Wertime, T. A., Eds., U.S. Government Printing Office, Washington (1978).

Chevallier, R. *Roman Roads,* B. T. Batsford (1976).

Chiles, J. R. Spindletop, *American Heritage of Invention and Technology,* 3(1), p. 35 (1987).

Clausius, R. J. E. *Abhandlungen uber die mechanische Wärmtheorie* (1864–1867). Trans. by W. R. Browne as *Mechanical Theory of Heat,* Macmillan (1879).

Clayton, A. R. K. The Shrewsbury and Newport Canals, *Thomas Telford: Engineer,* Penfold, A., Ed., Telford (1980).

Coel, M. A silver streak, *American Heritage of Invention and Technology,* 2(3), p. 10 (1986).

Collett, J. *Powered Lifting Devices,* Unpublished paper prepared for the Land and Water Development Division, FAO (1980).

Columella, Lucius Junius Moderatus. *Hortuli commentarium,* Rome, S. Plannck (1488–90?). Dibner Collection, National Museum of American History, Washington.

Cowan, R. S. Less work for mother? *American Heritage of Invention and Technology,* 2(3), pp. 57–63 (1987).

Cromwell, L. et al. *Biomedical Instrumentation and Measurements,* Prentice-Hall (1980).

Crosby, S. M. *The Abbey of St. Denis, 455–1122,* Yale (1942).

Davidovits, J. Ancient and modern concretes: what is the real difference? *Concrete Int.* 9(1), pp. 23–29 (1987).

DeCamp, L. S. *The Ancient Engineers,* Ballantine (1963).

Defoe, D. *A Tour thro' the Whole Island of Great Britain,* 2 Vols., Peter Davis (1927).

Delebrück, R., *Die Drei Tempel in Forum Holitorium,* Rome (1903).

Deswarte, S. and Lamoine, B. *L'Architecture et les Ingenieurs,* Moniteur (1979).

DiFenzio, Sulla portala degli antichi acquedotti romani e determinazione della quinaria, *Giornale del genio civile,* Rome (1916).

Dilke, O. A. W. *The Roman Land Surveyors.* David and Charles (1971).

Dilke, O. A. W. Mathematics for the surveyor and architect, *Archaeol. Today,* 8(8), p. 31 (1987).

Diodorus, Siculus. *The Library of History,* Davis (1814).

Dobrovolny, J. S. *A General Outline of Engineering History and Western Civilization,* Stipes (1958).

Dorling Kindersley, Ltd., Eds. *Quest for the Past,* Reader's Digest Association, Montreal.

Doscher, T. M. Enhanced recovery of oil, *Am. Sci.,* 69(2), p. 193 (1981).

Doty, K. L. *Fundamentals of Microcomputer Architecture,* Matrix (1979).

Duckert, J. M. *A Short History of the United States,* The Edison Electric Institute (1980).

Emme, E. M. Perspectives on the history of flight in America, *Two Hundred Years of Flight in America,* AAA History Series, Vol. 1, p. 24 (1977).

Etchwerry, B. A. and Harding, S. T. *Irrigation Practice and Engineering,* 2 Vols., New York (1933).

Ezell, E. C. The heroic era of space flight, *Two Hundred Years of Flight in America,* AAA History Series, Vol. 1 (1977).

Extra-Terrestrial Materials Processing and Construction, NASA (1980).

Falconer, S. E. An early 'Dark Age' in the southern Levant — insights on the basis of complex society, Paper presented at the 1st Joint Archaeological Congress, Baltimore (1989).

Fagan B. M. *People of the Earth,* Little, Brown (1980).

Finch, J. K. *The Story of Engineering,* Doubleday-Anchor (1960).

Finley, J. A description of the patent chain bridge, *The Portfolio,* Vol. 3 (1810).

Flink, J. J. The path of least resistance, *American Heritage of Invention and Technology,* 5(2), p. 34 (1989).

Flink, J. J. Innovation in automotive technology, *Am. Sci.,* 75(2), pp. 151–161 (1985).

Fourth Report of the Royal Commission of Pollution of Rivers in Scotland (1867).

Forbes, R. J. *Man the Maker,* Henry Schuman (1950).

Forman, W. *Byzantium,* Golden (1983).

Friberg, J. Numbers and measures in the earliest written records, *Sci. Am.,* 17, pp. 110–118 (1965).

Frontinus, Sextus Julius. *"De aquae ductu,"* The Two Books on the Water Supply of the City of Rome, (97).

Garrison, E. G. Engineering puzzle solved, *Civil Engineering,* 54(6), ASCE (1982).

Gibbon, J. H., Jr. Application of a mechanical heart and lung apparatus to cardiac surgery, *Minnesota Med.,* (1954).

Gibb-Smith, H. *The Aeroplane, An Historical Survey of its Origins and Development,* Her Majesty's Stationery Office (1960).

Gille, B., Ed. *Histoire Générale des Techniques,* Partie Médiévale (1959).

Gille, B. *Engineers of the Renaissance,* MIT Press (1966).

Gimpel, J. *The Cathedral Builders,* Grove (1961).

Goddard, R. H. A method for reaching extreme altitudes, *Smithsonian Publ.,* No. 2540 (1919).

Goldstine, H. H. *The Computer,* Princeton Press.

Hamblin, D. J. *The First Cities,* Time-Life (1973).

Hambly, E. C. The North Sea Challenge, *Great Engineers,* Walker, D., Ed., Academy-St. Martins (1987).

Harding, H. Tunnels, in *The Works of Isambard Kingdom Brunel,* Pugsley, A., Ed., Cambridge (1976).

Harrison, G. R. Production in the Beaufort Sea — feasible by the mid-1980's, *Ocean Ind.,* 15(8), pp. 35–39 (1980).

Hart, I. *The Mechanical Investigations of Leonardo da Vinci,* London (1925).

Hazen, A. *Clean Water and How to Get It,* 2nd ed., John Wiley (1907).

Heath, I. *Byzantine Armies, 886–1118,* Osprey (1979).

Hetherington, P. *Byzantium, City of Gold, City of Faith,* Golden (1983).

Hero. *Di Herone Alessandrine. De gli automati,* G. Ponno (1589).

Heronis Mechanici Liber de Machinis Bellicis. F. Franiscium (1572).

Herold, J. C. *The Swiss Without Halos,* Greenwood Press (1948).

Hill, B. H. The older Parthenon, *Am. J. Archaeol.*, XVI, pp. 535–558 (1912).

Hill, D. *A History of Engineering in Classical and Medieval Times,* Open Court (1984).

Ifrah, G. *From One to Zero,* Viking (1985).

Kasir, D. S. *The Algebra of Omar Khayyám,* (1931).

Kelly, D., Ed. *The Handbook of Waco and McLennan Counties,* (n.d.).

Kemp, P. *The History of Ships,* Longmeadow (1988).

Kenyon, K. M. Ancient Jericho, *Hunters, Farmers, and Civilizations: Old World Archaeology,* Freeman (1979).

Kerker, M. Sadi Carnot and the steam engineers, *Isis,* Vol. 51 (1960).

Kerr, N. Welch castles, *Popular Achaeol.,* 5(1), pp. 3–8 (1983).

Kirby, R. S., Thington, S. W., Darling, A. D., and Kilgour, F. G. *Engineering in History,* McGraw-Hill (1956).

Kunç, S. Analyses of Ikiztepe Metal Artifacts, *Anatolian Studies,* XXXVI, pp. 99–101 (1986).

Ley, W. *Engineer's Dreams,* Viking (1954).

Livy, Titus Livius. *History of Rome,* I. M. Dent & Sons (1912–1924).

Loftin, L. V., Jr. *Quest for Performance,* NASA (1985).

Los Angeles Times, Sunday, March 19, 1978.

de Lumley, H. A paleolithic camp at Nice, *Sci. Am.,* 220, pp. 42–50 (1969).

Lloyd, S. *The Archaeology of Mesopotamia,* Thames and Hudson (1984).

Makintosh, A. R. The first electronic computer, *Phys. Today,* 40(5), pp. 25–32 (1987).

Malek, J. *In the Shadow of the Pyramids,* Golden (1986).

Maiteus, J. The Vandals: myths and facts about a Germanic tribe of the first millenium A.D., *Archaeological Approaches to Cultural Identity,* Sherman, S. J., Ed., University of Hyman (1989).

Mark, R. Structural experimentation in Gothic architecture, *Am. Sci.,* 66(3), p. 550 (1978).

Mark, R. *Mystery of the Master Builder,* Coronet Film & Video (1990).

Marsh, P. *The Robot Age,* Abacus (1982).

Martin, L. Ancient Greek warship sails again, *Benthosaurus,* No. 25, p. 2 (1989).

Martinez, O. *Panama Canal,* Gordon & Cremonesi (1978).

Mason, C. F. Sanitation in the Panama Canal Zone, *The Panama Canal,* McGraw-Hill (1916).

Masters, D. *German Jet Engines,* Jane's (1982).

Maugh, T. H. II. A metallurgical tale of irony, *Sci. News,* 215, p. 153 (1981).

McBride, J. F. Solid-propellant rocket system, *Exploring in Aerospace Rocketry,* NASA (1971).

McEvedy, C. *The Penguin Atlas of Ancient History,* Penguin (1967).

McEvedy, C. *The Penguin Atlas of Ancient History,* Penguin (1981).

McKay, E. A. Tunneling to New York, *American Heritage of Invention and Technology,* 4(2), pp. 22–31 (1988).

Mendelssohn, K. A scientist looks at the pyramids, *Am. Sci.,* (59), pp. 210–220 (1971).

Middleton, J. H. *The Remains of Ancient Rome,* London (1892).

Mitchell, J., Ed. Railroads of the future, *Random House Encyclopedia,* pp. 1704–1705 (1983).

Mitchell, W. P. The hydraulic hypothesis: a reappraisal, *Current Anthropology,* 14(5), pp. 532–534 (1973).

Mock, E. B. *The Architecture of Bridges,* Museum of Modern Art (1949).

Mumford, L. History: Neglected clue to technological change, *Technology and Culture,* II, p. 235 (1961).

Mumford, L. *Technics and Civilization,* Harcourt, Brace and World (1963).

NASA, The First Twenty-five Years, 1958-1983, National Aeronautics and Space Administration (1983).

Nastian, T. J. Prehistoric Copper Mining in Isle Royale National Park, Michigan, Museum of Anthropology, Ann Arbor: Univ. of Michigan (1969).

Nef, J. U. An early energy crisis and its consequences, *Civilization*, W. H. Freeman (1979).

Newell, S. *Introduction to Microcomputing,* Harper and Row (1982).

Newton, I. *Philosophie Naturalis Principa Mathematica,* John Streater (1967). Dibner Collection, National Museum of American History.

Nicks, O. N. *The Far Travelers,* NASA (1985).

Oberth, H. Die Rakete zu dem Planetraumen. Munich: Oldenburg Publishing (1923).

Osterwalder, C. *La Suisse Préhistorique,* Editions 24 heures.

Overskel, D. Advanced Tokamaks and a path to ignition, *Phys. Today,* 41(1), 1988.

Parsons, W. B. *Engineers and Engineering in the Renaissance,* MIT Press (1939).

Paxton, R. A. Menai Bridge, 1816–26, *Thomas Telford: Engineer,* Thomas Telford, Ltd. (1980).

Penfold, A. Managerial organization on the Caledonian Canal, *Thomas Telford: Engineer,* Penfold, A., Ed., Telford (1980).

Peters, T. F. *Transitions in Engineering,* Birkhauser (1987).

Pevsner, N. *An Outline of European Architecture,* Pelican (1968).

Philon of Byzantium. *Le Livres des Appareils Pneumatiques et des Machines Hydrauliques par Philon de Byzance,* in Notices et Extraits des Manuscrits de la Bibliothèque Nationale, Tome 38 (1903).

Philon of Byzantium. *Pneumatics,* LXIII, Ed. Cana de Vaux, Paris Académie des Inscriptions et Belle Lettres, 38 (1903).

Phonecia, *The American International Encyclopedia,* The John C. Winston Company (1954).

Pope, T. *A Treatise on Bridge Architecture,* A. Niven (1811).

Prager, F. D. and Scaglia, G. *Mariano Taccola and His Book DE INGENEIS,* MIT Press (1972).

Prager, F. D. A manuscript of taccola quoting Brunelleschi, on problems of inventors and builders, *Proc. Am. Philos. Soc.* CXII, pp. 131–149 (1968).

Procopius. *Secret History,* Time (1966).

Redman, C. L. *The Rise of Civilization,* W. H. Freeman (1978).

Renfrew, C. Recalibrated radiocarbon dates place the final construction date at 1800 B. C., *The Emergence of Civilization,* Alfred A. Knopf (1975).

Rivoira, G. T. *Roman Architecture,* Oxford (1925).

Rollefson, G. and Simmons, A. H. The life and death of 'Ain Ghazal, *Archaeology,* 40(6), pp. 38–45 (1987).

Robinson, R. BULLWINKLE, *Civil Eng.,* 59(7), pp. 34–37 (1989).

Russell, G. E. *Hydraulics,* Henry Holt (1937).

Sackheim, D. E. *Historic American Engineering Record Catalog,* U.S. Government Printing Office (1976).

Sandström, G. E., *Man the Builder,* McGraw-Hill (1970).

Schwarzchild, B. Princeton Tokamak reaches record plasma ion temperature, *Phys. Today,* 39(11), 1986.

Science Digest, February 1983.

Sherrad, P. *Byzantium,* Translated by Richard Atwater. Michigan (1961).

Skinner, S. K. and Reilly, W. K. The EXXON VALDEZ Oil Spill, *Oil Spill Intelligence Report* May (1989).

Space Shuttle, NASA (1975).

Shiffman, M. A. Food and environmental sanitation, in *History of Environmental Sciences and Engineering,* School of Public Health, University of North Carolina at Chapel Hill, Department of Env. Sci. and Eng. Pub. No. 259 (1971).

Schutz, H. *The Prehistory of Germanic Europe,* Yale (1983).

Smith, N. A. F. *Man and Water,* Peter Davies (1971).

Smith, N. A. F. *A History of Dams,* Peter Davies (1971).

Snow, J. A reprint of two papers by John Snow, M. D., The Commonwealth Fund, New York 1855.

Sekler, E. F. *Wren and His Place in European Architecture,* Macmillan (1956).

Seguin, M. *Des Ponts en Fil de Fer par Seguin Aine d'Annonay,* Bachelier (1824).

Smeaton, J. *A Narrative of the Building and a Description of the Eddystone Lighthouse with Stone,* London (1791).

Steward, J. H., Ed. *Irrigation Civilization, A Comparative Study,* Pan American Union (1955).

Stüssi, F. *Schweizerische Bauzeitung,* Vol. 116 (1940).

Stranges, A. N. Synthetic petroleum from high pressure coal hydrogenation, *Chemistry and Modern Society,* American Chemical Society (1983).

Straub, H. *A History of Civil Engineering,* The MIT Press (1964).

Tarkov, J. Engineering the Erie Canal, *American Heritage of Invention and Technology,* 2(1), pp. 50–57 (1986).

Taub, A. H., Ed. *Collected Works of John von Neumann,* Vol. 5, Macmillan (1963).

Taylor, R. A mechanical heart-lung apparatus, *IBM J. Res. Dev.,* 1, p. 300 (1951).

Thulin, C. Die Handschriften deu Coprus Agrimensorum Romanorum. *Abhand Königlich Pressische Akademie der Wiss. Philos. Hist.,* Classe, Part II (1911).

Trattato dei pondi, leve e triari. (Codice Laurenziano, No. 361, Serie Ashburnham), now in the Laurentian Library at Florence.

Trinder, B. The Holyhead Road: an engineering project in this social context, *Thomas Telford: Engineer,* Penfold, A., Ed., Telford (1980).

Van Den Broucke, S. Aqueducts, cisterns, and siphons, *Archaeol. Today,* 8(4), p. 18 (1987).

Vander Merwe, N. J. and Avery, D. H. Pathways to steel, *Am. Sci.,* 70(2), pp. 146–155 (1982).

Vicat, L. J. *Observations Diverses sur la Force et al Duree des Cables de Fer,* APC (1836).

da Vinci, Leonardo. *Leonardo da Vinci's Notebooks English Arranged and Rendered Into English by Edward McAudy,* New York (1923).

Viollet-Le-Duc, E. E. *Dictionnaire Raissoné de Mobilier Français de l'Epoque Carlovingiennee à la Renaissance,* 6 Vols., Gründ et Maguet (1914).

Vitruvius, Marcus Pollio, *De architectura libri decem.*

Vitruvius, M. P. *De Architectura,* 2 vols., Loeb Classics, London (1970).

Vladimirov, L. *The Russian Space Bluff,* London (1971).

Walker, D. The Railway engineers, *Great Engineers,* Walker, D., Ed., Academy-St. Martins (1987).

Wells, G. M. Municipal engineering and domestic water supply in the canal zone, *The Panama Canal,* McGraw-Hill (1916).

Wells, H. G. *The Outline of History,* Vol. 1, Doubleday (1971).

Wertime, T. A. The search for ancient tin, the geographic and historic boundaries, in *The Search for Ancient Tin,* Franklin A. D., Olin, J. S., and Wertime, T. A., Eds., U.S. Government Printing Office (1978).

Winter, F. H. *The Rocket Societies: 1924–1940,* Smithsonian (1983).

Winter, H. J. J. *Eastern Science,* John Murray (1952).

Wittfogel, K. A. *Oriental Despotism: A Comparative Study of Total Power,* Yale University Press (1957).

Wood, M. *In Search of the Trojan War,* Facts on File Video (1985).

Woodcraft, B., Ed. *The Pneumatics of Hero of Alexandria,* Walton and Maberly (1951).

INDEX

H